Advances in Intelligent Systems and Computing

Volume 603

Series editor

Janusz Kacprzyk, Polish Academy of Sciences, Warsaw, Poland
e-mail: kacprzyk@ibspan.waw.pl

About this Series

The series "Advances in Intelligent Systems and Computing" contains publications on theory, applications, and design methods of Intelligent Systems and Intelligent Computing. Virtually all disciplines such as engineering, natural sciences, computer and information science, ICT, economics, business, e-commerce, environment, healthcare, life science are covered. The list of topics spans all the areas of modern intelligent systems and computing.

The publications within "Advances in Intelligent Systems and Computing" are primarily textbooks and proceedings of important conferences, symposia and congresses. They cover significant recent developments in the field, both of a foundational and applicable character. An important characteristic feature of the series is the short publication time and world-wide distribution. This permits a rapid and broad dissemination of research results.

More information about this series at http://www.springer.com/series/11156

Tareq Ahram
Editor

Advances in Human Factors in Sports, Injury Prevention and Outdoor Recreation

Proceedings of the AHFE 2017 International Conference on Human Factors in Sports, Injury Prevention and Outdoor Recreation, July 17–21, 2017, The Westin Bonaventure Hotel, Los Angeles, California, USA

Springer

Editor
Tareq Ahram
Institute for Advanced Systems Engineering
University of Central Florida
Orlando, FL
USA

ISSN 2194-5357 ISSN 2194-5365 (electronic)
Advances in Intelligent Systems and Computing
ISBN 978-3-319-60821-1 ISBN 978-3-319-60822-8 (eBook)
DOI 10.1007/978-3-319-60822-8

Library of Congress Control Number: 2017943055

Printed on acid-free paper

This Springer imprint is published by Springer Nature
The registered company is Springer International Publishing AG
The registered company address is: Gewerbestrasse 11, 6330 Cham, Switzerland

Advances in Human Factors and Ergonomics 2017

AHFE International
Applied Human Factors and Ergonomics Conference
2017
www.ahfe.org

AHFE 2017 Series Editors

Tareq Z. Ahram, Florida, USA
Waldemar Karwowski, Florida, USA

8th International Conference on Applied Human Factors and Ergonomics and the Affiliated Conferences

Proceedings of the AHFE 2017 International Conference on Human Factors in Sports, Injury Prevention and Outdoor Recreation, July 17–21, 2017, The Westin Bonaventure Hotel, Los Angeles, California, USA

Advances in Affective and Pleasurable Design	*WonJoon Chung and Cliff (Sungsoo) Shin*
Advances in Neuroergonomics and Cognitive Engineering	*Carryl Baldwin*
Advances in Design for Inclusion	*Giuseppe Di Bucchianico and Pete Kercher*
Advances in Ergonomics in Design	*Francisco Rebelo and Marcelo Soares*
Advances in Human Error, Reliability, Resilience, and Performance	*Ronald L. Boring*
Advances in Human Factors and Ergonomics in Healthcare and Medical Devices	*Vincent G. Duffy and Nancy Lightner*
Advances in Human Factors in Simulation and Modeling	*Daniel N. Cassenti*
Advances in Human Factors and System Interactions	*Isabel L. Nunes*
Advances in Human Factors in Cybersecurity	*Denise Nicholson*
Advances in Human Factors, Business Management and Leadership	*Jussi Kantola, Tibor Barath and Salman Nazir*
Advances in Human Factors in Robots and Unmanned Systems	*Jessie Chen*
Advances in Human Factors in Training, Education, and Learning Sciences	*Terence Andre*
Advances in Human Aspects of Transportation	*Neville A. Stanton*

(continued)

(continued)

Advances in Human Factors, Software, and Systems Engineering	*Tareq Z. Ahram and Waldemar Karwowski*
Advances in Human Factors in Energy: Oil, Gas, Nuclear and Electric Power Industries	*Paul Fechtelkotter and Michael Legatt*
Advances in Human Factors, Sustainable Urban Planning and Infrastructure	*Jerzy Charytonowicz*
Advances in the Human Side of Service Engineering	*Louis E. Freund and Wojciech Cellary*
Advances in Physical Ergonomics and Human Factors	*Ravindra Goonetilleke and Waldemar Karwowski*
Advances in Human Factors in Sports, Injury Prevention and Outdoor Recreation	*Tareq Z. Ahram*
Advances in Safety Management and Human Factors	*Pedro Arezes*
Advances in Social & Occupational Ergonomics	*Richard Goossens*
Advances in Ergonomics of Manufacturing: Managing the Enterprise of the Future	*Stefan Trzcielinski*
Advances in Usability and User Experience	*Tareq Ahram and Christianne Falcão*
Advances in Human Factors in Wearable Technologies and Game Design	*Tareq Ahram and Christianne Falcão*
Advances in Communication of Design	*Amic G. Ho*
Advances in Cross-Cultural Decision Making	*Mark Hoffman*

Preface

Human Factors in Sports, Injury Prevention and Outdoor Recreation aims to address the critical cognitive and physical tasks which are performed within a dynamic, complex, collaborative system comprising multiple humans and artifacts, under pressurized, complex, and rapidly changing conditions that take place during the course of any sporting event. Highly skilled, well-trained individuals walk a fine line between task success and failure, with only marginally inadequate task execution leading to loss of the sport event or competition. This conference promotes cross-disciplinary interaction between the human factors in sport and outdoor recreation disciplines and provides practical guidance on a range of methods for describing, representing, and evaluating human, team, and system performance in sports domains. Traditionally, the application of human factors and ergonomics in sports has focused on the biomechanical, physiological, environmental, and equipment-related aspects of sports performance. However, various human factors methods, applied historically in the complex safety critical domains, are suited to describing and understanding sports performance. The conference track welcomes research on cognitive and social human factors in addition to the application of physiological ergonomics approaches sets it apart from other research areas. This book will be of special value to a large variety of professionals, researchers, and students in the broad field of Sports and Outdoor Recreation. Three sections presented in this book are as follows:

I. Injury Prevention and Analysis of Individual and Team Sports
II. Physical Fitness and Exercise
III. Assessment and Effectiveness in Sports and Outdoor Recreation

Each section contains research that has been reviewed by members of the International Editorial Board. Our sincere thanks and appreciation to the Board members as listed below:

Damian Morgan, Australia
Timothy Neville, Australia
E. Salas, USA
Daniel Simmons, UK
Neville Stanton, UK
Scott Talpey, Australia
Guy Walker, UK
P. Waterson, UK

This book will be of special value to a large variety of professionals, researchers, and students in the field of performance who are interested in Injury and Accidents prevention, and design for special populations, particularly athletes. We hope this book is informative, but even more that it is thought provoking. We hope it inspires, leading the reader to contemplate other questions, applications, and potential solutions in creating good designs for all.

July 2017 Tareq Ahram

Contents

Assessment and Effectiveness in Sports and Outdoor Recreation

Effect of Rater Expertise on Subjective Agility Assessment

Chika Eke(✉) and Leia Stirling

Massachusetts Institute of Technology, Cambridge, MA, USA
{ceke, leia}@mit.edu

Abstract. Agility performance is often quantified using completion time, which provides little information about which factors contribute to or limit an individual's performance. The objective of this study was to determine how novices and experts working in athletic, clinical, and military environments qualitatively and quantitatively evaluate agility performance. Formalizing expert definitions will inform the development of objective biomechanical metrics, which have the potential to inform strategy development for training and rehabilitation. Thirty-three participants completed a survey which involved scoring 16 athletes on a 7 point Likert scale of not agile to agile. The spread of the scores indicated that even within groups, participants had different opinions about which aspects of technique contributed to high performance. Participant responses were used to link several terms to agility technique. Future work includes quantitatively defining and evaluating these terms.

Keywords: Human factors · Performance assessment · Agility

1 Introduction

A common definition of agility is the ability to quickly change speed or direction [1]. Two types of agility are discussed in literature—planned agility and reactive agility. Planned agility includes a course that requires the physical act of changing direction, where the person knows the course *a priori* and navigates a predefined path. Reactive, or unplanned agility, incorporates a cognitive component by involving perception and reaction to an external cue [2]. For reactive agility, the course is not pre-planned and direction changes are signaled during the navigation of the course. It is well established that the ability to change direction is an important performance variable for predicting success in field sports. Multiple planned agility tests have been implemented for evaluation purposes [3, 4]. Three of the most commonly used tests are the T-Test, Illinois Agility Test, and 505 Test. The T-Test, named for the shape of the associated course, requires 4 directional changes. The athlete runs from the start line to a cone approximately 10 m ahead, side steps to a cone 5 m to the left of the center cone, side steps in the opposite direction to a cone 5 m to right of the center cone, sidesteps from the right cone to the center once again, and backpedals to the start line [5]. The Illinois Agility Test is a timed task involving straight sprinting and weaving through 4 cones. The movement patterns resemble those applied to dodge opponents in soccer and rugby [6]. To complete the 505 test, which was originally designed for cricket players,

T. Ahram (ed.), *Advances in Human Factors in Sports, Injury Prevention and Outdoor Recreation*, Advances in Intelligent Systems and Computing 603,
DOI 10.1007/978-3-319-60822-8_1

athletes sprint 5 m forward from a start line, pivot 180 degrees and return to the start line [7]. Although these tests accurately replicate the sharp direction changes required in multiple athletic environments, they do not address the cognitive processes contributing to swift movements when reacting to an opponent.

A few studies have addressed the cognitive aspects of agility by developing tests with unpredictable stimuli. Spasic [8] designed a course, similar to the T-Test, for handball players that required participants to react to visual cues. LEDs placed within one of two cones lit up in a randomized order each time the participant crossed an infrared beam during the straight sprint. Athletes had to assess which cone was illuminated and shuffle to that cone as quickly as possible. A perceptual-reactive-capacity index (the ratio of completion time for the reactive version of the course divided by completion time of the planned version of the course) was examined with the hypothesis that it would differentiate between defensive and offensive handball players. The study supported the hypothesis that defensive players, who regularly react to opponents' actions, having a better perceptual-reactive-capacity index than offensive players, who primarily perform planned changes in direction.

Other reactive agility studies have assessed anticipation skills and decision time using stimuli provided in real-time by another person or through a video clip of an athlete performing a set of sport-specific movements [2, 9]. Sekulic et al. [10] developed an agility course that permitted evaluation of variation in cutting angle, while enabling flexibility in running technique (side stepping not required), incorporation of an external cue, and was unique from other courses by requiring athletes to come to an abrupt stop and accelerate out of breakpoints. Performance time in this course differentiated between college-aged athletes involved in agility-saturated sports (soccer, basketball, handball, volleyball) and those not involved in agility-saturated sports (gymnastics, dance) [10].

The planned and reactive agility tests typically quantify agility performance using time-based metrics—primarily the time elapsed between crossing the start and finish line. While speed is important for agility, the parameter does not provide insights about strategy or technique, which enable identifying areas of improvement and risk of injury. These insights on technique are typically obtained from experts that visually assess agility tasks qualitatively. Previous studies have examined particular components of technique (e.g., straight sprinting performance, leg strength, and power qualities evaluated by jumping tasks) and found weak correlations to overall agility course time [11–14]. Evaluation of biomechanical data has found a subset of parameters that were sensitive to a sharp change of direction (e.g., trunk flexion, ground contact time, ankle power, ankle plantar flexor moment, and knee flexion) [15, 16]. These studies highlight that there are potential measures that may inform on technique, but they still rely on cutting time as the predictor for optimal performance. It is unclear from the literature whether additional measures should be considered beyond speed for assessing agility performance and how experts qualitatively make decisions on agility performance.

The objective of this study was to determine how experts evaluate agility and to identify key terms defining agility performance. The metrics identified will enable a focused examination of new parameters for assessing agility technique and will extend previous studies that have found weak correlations when comparing to solely course time, enabling the identification of performance strengths and weaknesses. The quantification of methods for assessing technique can lead to objective evaluations that can be completed by non-experts. While many evaluations in the literature consider sports performance, agility tasks are also relevant in service member training and rehabilitation for movement disorders [17, 18]. Here we specifically consider how agility is characterized by athletic, clinical, and military experts when viewing the same task and group of participants. The task selected for the user groups to evaluate was athletes performing the reactive agility task defined by Sekulic [10]. This type of comparison is useful for understanding the invariant components within agility and how quantified parameters may be generalized across domains. Variations in environment and performance expectations for each area of expertise may drive differences in qualitative assessment. For example, a physical therapist may place less emphasis on speed than a soccer coach, given a desire for patients to develop healthy movement patterns rather than react quickly to an external cue. Further, we anticipate that even though all experts were trained in their discipline, there may be variability within as well as across disciplines based on different specialties or sub-specialties.

To extend the understanding of agility performance beyond speed-based measures, this study investigated how athletes with comparable speeds were ranked. Rankings using internal reference frames (a Likert score) and forced reference frames (explicit ranks) were considered. Maio et al. [19] discussed the potential differences between the two, highlighting that rankings of ethical acceptability of behaviors using scores were more correlated with *a priori* predictions than explicit ranks. The investigators argued that explicit rankings may cause participants to make unimportant distinctions that would not have been made otherwise. However, the additional distinctions explicit rankings may generate by forcing participants to be more detail-oriented may be particularly useful for assessing human performance. We included both ranking methods in order to further evaluate these relationships.

In this study, we hypothesize that (1) the definition of agility differs by expert background; (2) assessments within group are similar; and (3) the rankings assessed through a forced reference frame differ from an internal reference frame. To consider the consistency of the internal reference frame, we have the scorers view the same athlete twice and we assess the additional hypothesis that (4) scores are consistent between viewings of the same athlete.

2 Methods

2.1 Participants

The study was completed by 33 adults (mean age 30 years, SD = 9 years; 16 female). Participants were recruited within an expert group—athletic (n = 8), clinical (n = 7), military (n = 8)—or novice group (n = 10) based on their experience evaluating human performance. Expert groups were familiar with formal training and evaluation guidelines within their field. The novice group had no previous knowledge of formal guidelines. The athletic group consisted of coaches specializing in football, rugby, soccer, field hockey, tennis, and track. The clinical group consisted of physical therapists. The military group included experienced members of Air Force and Army Reserve Officers' Training Corps (ROTC).

2.2 Athlete Videos

The videos analyzed within this user study by the expert and novice participants were obtained from a previously collected data set. The reactive agility obstacle (Fig. 1) was a sub-set of the obstacles performed by the athletes. To complete the obstacle, athletes (n = 16) ran from the start line to an endpoint, touched the top of the endpoint cone, ran back to the start line, and turned around to repeat these actions for three more endpoints as quickly as possible. Endpoints were vocally announced each time the athletes crossed the cue line. Athletes were not provided a strategy on how to complete the task. Half of the athletes completed the reactive agility obstacle 6 times, while the other half completed this obstacle 3 times. All athletes provided written consent and procedures were approved by the University of Michigan IRB and the MIT Committee on the Use of Humans as Experimental Subjects (COUHES). Athletes were compensated up to $50 for their participation. The videos were parsed and the reactive agility videos of the athletes on their first two times through the obstacle were used within the user study. Videos were de-identified by blurring participant faces using Adobe After Effects software. Athlete videos were categorized as slow, medium, or fast groups based on the time it took them to complete the course. Videos were shown at real-speed and not normalized for time.

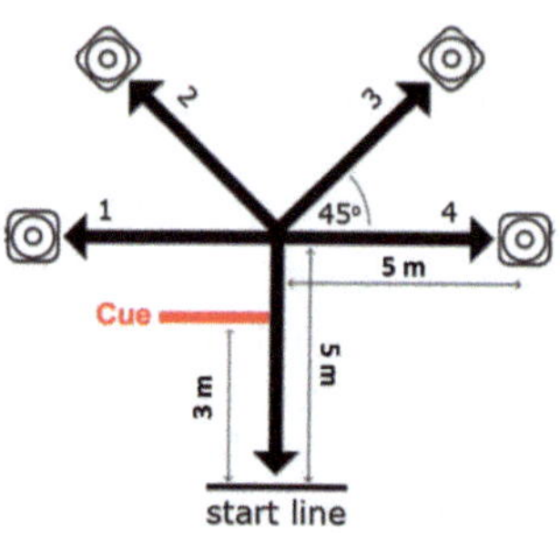

Fig. 1. Reactive agility course adapted from Sekulic [10]. Athletes received verbal cues at the location notated and touched 4 endpoint cones per trial.

2.3 User Study Experimental Protocol

Procedures for the user study were approved by the MIT COUHES and all participants provided written consent. Participants received up to $20 in compensation. Participants completed an online agility evaluation survey consisting of 4 parts. Part 1 was a short answer question asking for any terms or definitions that the participant associated with agility performance. Part 2 presented the videos showing the 16 athletes completing their second time through the reactive agility course (Sect. 2.2). Participants were asked to score each athlete's video on a Likert scale ranging from 1 (not agile) to 7 (very agile). Each video was approximately 45 s long and was presented on a new page of the form in a randomized order. Participants took a 10 min break after the first 16 videos. The second set of 16 videos showed the athletes completing their third time through the reactive agility course and were presented in mirrored order, without informing participants of the repetition of athletes. There was an option to take a 5 min break before beginning Part 3 of the survey, which requested a ranking of agility performance. Two sub-sets of 5 videos from the group of 16 athletes were arranged on the same page and participants ranked each set of videos from most agile to least agile. Both sub-sets contained a mixture of videos from the first and second set of athlete videos. The first sub-set of 5 videos included the performance of 1 fast and 4 medium speed athletes. The second sub-set contained 1 fast, 1 medium, and 3 slow athletes. Both the scoring and ranking sections of the survey prompted participants to provide explanations for their selections. Part 4 of the survey provided space for further explanation if the participant's definition of agility had changed based on watching the videos. Survey completion time ranged from 1 to 2 h.

2.4 Data Analysis

A Wilcoxon Signed Rank test was used to evaluate difference in rater score between first and second videos for the athletes. A paired t-test was used to assess difference in course completion time between the first and second videos for the athletes. A Kruskal-Wallis test was used to evaluate differences in score between groups. Differences between rankings as determined through scores and explicit ranks were determined with a Chi-squared test. The fourth spread of the scores was calculated for each video to quantify variability. This calculation involved ordering the observations of data from smallest to largest and subtracting the median of the lower half of the data from the median of the upper half of the data. The fourth spread was chosen as an alternative to standard deviation because of its use of median values instead of mean values, which is more appropriate for Likert scale data [20].

A qualitative analysis was performed to identify the most common descriptors for agility performance. An initial list of terms to describe commonly used phrases in the survey explanations was developed by a first pass through of the qualitative data. Subsequent passes through all terms was made to assess if a phrase by a rater aligned with a term, or if a new term needed to be generated. Similar terms or phrases were combined and the coding scheme was refined upon follow-on passes through the terms. Frequencies for each term were assessed as the number of participants who used it.

3 Results and Discussion

3.1 Analysis of Qualitative Descriptions

The survey responses (Table 1) demonstrated that participants evaluated agility most frequently using terms related to athlete speed and ability to change direction, which aligns with the definition of agility found in literature [1]. Examples of phrases coded as speed and change of direction were "time through the course" and "sharp

Table 1. Agility terms

Term	Example phrase	Frequency
Speed	Quickness, foot speed and time through the course	30
Change direction	Cutting, pivoting	24
Efficient path	Arcing paths, distance from cone on turns	23
Reaction time	Good reflexes, responds to commands in timely manner	21
Body alignment	Lowering center of gravity in and out of numbered breakpoints, bends well at the knees giving her sharpness changing direction	20
Acceleration	Quick starts and stops, acceleration out of turns	13
Foot contacts	Unnecessary steps before breakpoints, double footed turns, long foot contacts	13
Arm motion	She is not using her arms fully, can use arms more to pump	11
Smooth	Very smooth runner, fluid movements	7
Coordination	Disjointed, legs trunk and arms all coordinated in the position changes	6
Stride	Long strides and at a good speed, shorter stride length and accurate change of direction	6

movements when cutting and turning." The next frequently used term, "efficient path" is closely tied to the ability to change direction. Several raters commented that an athlete's ability to cut his or her body "quickly in the given direction without requiring any arcing paths to get there" was important. The efficient path term is distinct from the change of direction term as it highlights a particular strategy for making the turn, specifically the ability make precise turns towards the desired endpoint by minimizing path length. The high frequency of performance speed was supplemented by the term "reaction time," which is a focus on the response time after cue calls. Experts repeatedly mentioned decision-making in their responses, which highlights the importance of cognitive performance in the agility task. Their comments align with the agility definition provided by researchers such as Spiteri et al. [2], which discuss the correct identification and rapid interpretation of environmental cues in addition to changing direction. Another term that emerged from the survey responses was "body alignment," which included comments such as lowering the center of mass while bending at the knee and hip joints. Participants suggested that a proper body alignment enabled athletes to make sharp changes in direction, burst out of the course's break-points, and decelerate with full control. While related to speed and direction change, acceleration was categorized as a separate term as locations within the course could be performed using a constant speed direction change. Expert comments related to acceleration during the course provided additional information on strategy. Foot contacts provide additional information on athlete technique, with a given body speed having the potential for few or many contacts. Experts noted that athletes with good footwork minimized the amount of steps taken to make a turn and used "short, quick steps" or "good stutter stepping." They also mentioned that tight pumping arm motions aided athletes in changing direction and maintaining stability. Those that did not adequately pump their arms appeared to be less energetic. A smaller frequency of participants mentioned the value of making smooth movements, which may contradict with the stutter stepping strategy, efficient path, and abrupt body movements contributing to quick changes in direction.

In the last section of the survey, participants were asked to discuss whether the definition they provided for agility at the beginning of the survey had changed after viewing the videos. While many novices explicitly noted they adapted their definition ($n = 8$ of 10), fewer participants made this explicit assessment in the expert groups ($n = 3$ out of 8 athletic experts, $n = 3$ out of 7 clinical experts, and $n = 2$ out of 8 military experts). It was expected that novice definitions would experience the most change given their lack of exposure to formal agility evaluation methods. Some experts commented that while their general view of agility remained the same, the factors they considered to contribute to this view were dependent on the selected drill and were easier to articulate after reviewing the videos. For example, one expert in the athletic group expanded on his initial listing of speed and body control at the start of the survey to include "sharp, quick turns with the subject accelerating out of the turn using their arms". Other experts mentioned a new consideration of "bend in the knee and hip to allow twist and drive" to quantify readiness as well as the "accuracy of movement pathway."

3.2 Effect of Viewing Number on the Agility Score

Higher scores were provided by the clinical ($p < .01$), military ($p < .01$), and novice ($p < .05$) groups for the second set of videos than for the first set (Fig. 2). This result does not support Hypothesis 4, that scores would remain consistent during both evaluations of the same athlete. There was no significant difference in athlete time through the course for the two videos shown in the survey ($p = .282$). Differences in scoring may be due to participants having been unable to gauge the range of athletic skillset in performance before beginning the survey and therefore they relied on an internal representation of performance. Clinical, military, and novice groups may have adjusted their internal reference after the first set of viewings. The updated clarity in definition mentioned by participants at the end of the survey (Sect. 3.1) aligns with the difference in Score 2 observed for some groups. As the selected reactive agility task was from the athletic literature, there is a possibility that the athletic group was more familiar with assessing agility with similar tasks, creating a more informed initial representation that was not adjusted to a significant level. This difference in scoring for some groups informed the decision to assess within and across group differences using Score 2 for further analysis.

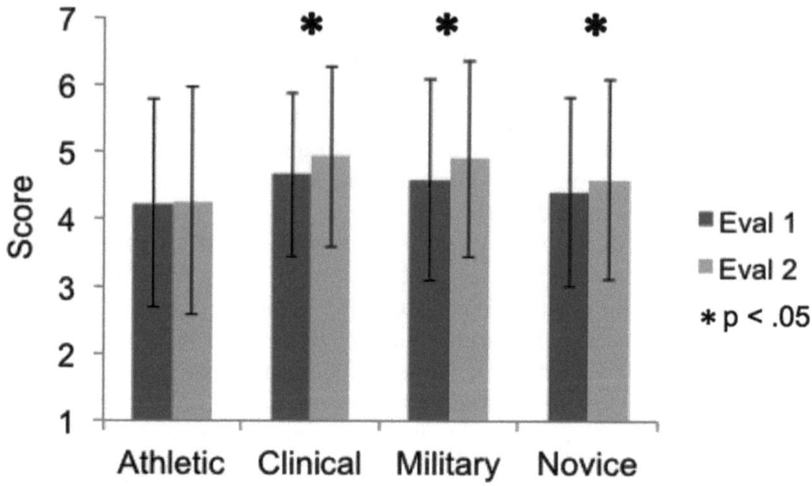

Fig. 2. Average group scores for first and second video evaluation. Scores ranged from 1 (low agility) to 7 (high agility). The asterisks (*) represent Wilcoxon Signed Rank test results with p-values below .05.

3.3 Effect of Expertise on the Agility Score

Score 2 was only significantly different between groups for one video (Video 2, $p < .05$) (Fig. 3). This outcome does not support Hypothesis 1, which states that the definition of agility differs by expert background. What was observed was variability even within groups. The scoring disagreement between groups for Video 2 stemmed from the athlete's good technique but slow pace, according to the scoring explanations provided by the participants. While speed was one of the most popular metrics considered to contribute to agility (see Table 1), some groups gave more weight to metrics related to strategy. The clinical group prioritized metrics that were independent of speed such as efficient turns and skillful footwork to cut in the proper direction. Conversely, most evaluators in the athletic group heavily penalized the performance for low speed.

Trends from Fig. 3 indicate that videos 6, 10, and 12 received the highest median scores from each group. The comments made by participants for these videos were in agreement about fast pace and good technique contributing to high performance. Participants specifically mentioned that these 3 athletes had fast reaction times, made quick turns, and lowered their center of gravity to touch the cones.

The spread of responses within groups fluctuated by video presented and was largest for the athletic and novice groups (Fig. 4). The spread in novice users is likely a result of individuals without basic training with which to guide their evaluations. However, the results for the athletic group do not support Hypothesis 2. While athletic-driven agility courses are used across multiple sports, individual sports may still value different components of agility performance. The variation in athletic group scoring may arise from our inclusion of a variety of sports. For example, the athletic group consisted of coaches from sports such as such as soccer and tennis, which differer

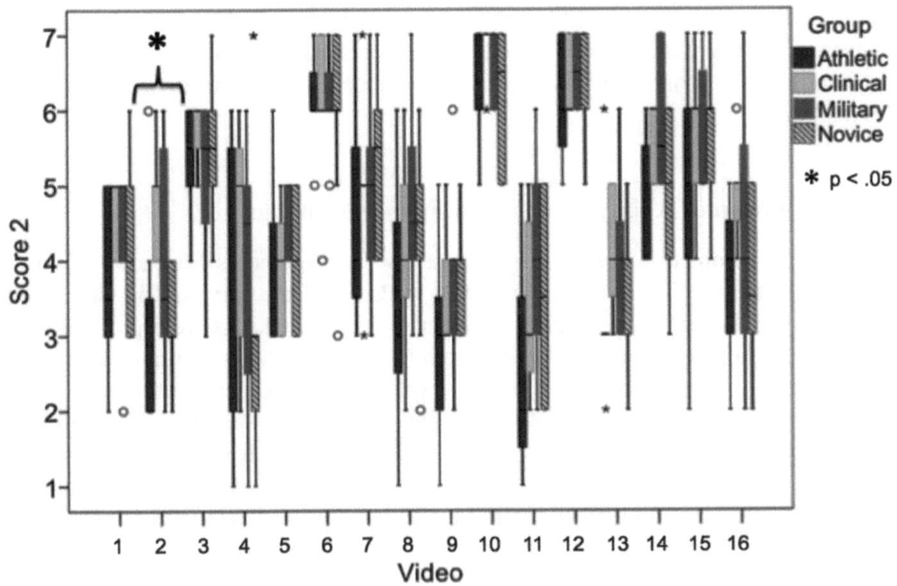

Fig. 3. Score 2 distribution between groups. Scores ranged from 1 (low agility) to 7 (high agility). The asterisks (*) represent Kruskal-Wallis test results with p-values below .05.

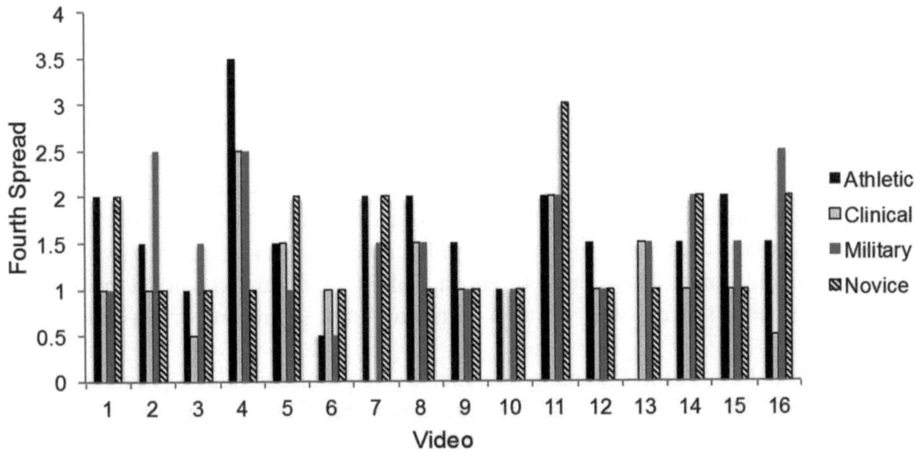

Fig. 4. Fourth spread of score 2 within groups for each observed video.

in required skillset. Large fourth spreads were observed across most groups for the evaluation of videos 4 and 11. Participants commented that the athletes in these videos were fast but had poor technique. There were disagreements within groups about what constituted poor technique, with some evaluators mentioning poor posture, while others discussed slow decision making and a lack of coordination. This variability in responses implies that even within groups, participants had different opinions about which aspects of technique contributed to high performance. Additionally, the variability in the rating of a fast athlete indicates that speed alone does not make an individual agile.

3.4 Comparison of Agility Scores with Forced Rankings

Hypothesis 3 examined if the ranking created by pooling scores for each athlete was different from the explicit ranking completed in Part 3 of the survey. The Chi-squared test results revealed that significantly different ($p < .01$) rankings were provided for 4 out of the 10 athletes evaluated using both methods. It is important to note that participants were forced to give different explicit ranks for each athlete while the scoring section of the survey permitted ties. The difference in ranking procedure is one possible source of variability in these two ordering methods. It was also observed that the 4 athletes with different rankings were either classified as medium speed out of the possible fast, medium, and slow categories or had speeds that were approximately equal to other athletes with which they were ranked. In these cases, the difficulty in discerning performance by speed alone likely drove participants to consider technique in ways that may not have been considered when scoring athletes individually. The forced rankings provide additional support that participants had varying internal valuations on the metrics for evaluating athletes.

4 Conclusion

The objective of this study was to determine how experts evaluate agility and to identify key terms defining agility performance. The metrics identified have potential to aid in quantifying agility for training and rehabilitation in clinical, military, and athletic environments. The survey analysis found that expert decision-making is guided by technique-based metrics in addition to speed-based metrics. These findings are based on qualitative analysis of the participant-provided descriptions and quantitative analysis of the scores and ranks. The value placed on certain strategies was not dependent on area of expertise as scoring was variable within and across groups for several athletes scored.

There are important limitations to consider in the presented study. A larger sample size may have aided in accommodating subgroups within the expert groups. Subgroups would have prevented the pooling of sub-specializations, which may look for different skillsets, and may have reduced the variability observed within groups. Another limitation of the survey was the use of videos filmed from inconsistent angles, which some participants stated made athletes appear faster or slower. While a forced ranking across all videos would have been interesting to examine, sub-set rankings with representative selections met the goal of identifying whether participants used technique to differentiate athletes with similar performance times.

The qualitative analysis summarizing the agility techniques noted by the participants can be used to define quantitative biomechanical metrics. There is opportunity to select metrics that are possible to robustly estimate using mocap, as well as defining measures that map to these terms using data from wearable sensors. The use of wearable sensing enables data collection in a natural setting, which extends the tasks and environments that may be assessed. The definitions of quantitative metrics that map to the qualitative terms provide a means to examine the multiple components that combine to enable an interpretation of agility. Similar to a decision-maker, these component metrics could be combined to construct a composite agility score. For example, the composite could be defined as a weighted average, with the frequency with which terms occurred in the survey used to define the weights. However, the variability in responses for individuals within and between groups highlights that such a composite may need to be tuned to address the strategies desired by a particular user or have weightings shown explicitly so that it can be interpreted by users who prioritize different techniques. The development of quantitative scores will enable a better understanding of a person's strategy and can aid in detecting areas for performance training beyond the time-based methods currently used. These methods will also be valuable in assessing operational decisions for military environments, or rehabilitation needs in a clinical environment. For example, quantitative scores could inform how selected military gear affects agility and could aid clinicians in selecting a plan of care using metric-based patient progress.

Acknowledgments. This work was supported by the US Army Natick Soldier Research, Development and Engineering Center (W911QY-13-C-0011) and a National Science Foundation Graduate Research Fellowship under Grant No. 1122374. Any opinions, findings, and conclusions or recommendations expressed in this material are those of the author(s) and do not necessarily reflect the views of the sponsor. The author's would like to thank our collaborators at the University of Michigan for providing the videos of the athletes on the obstacle course and performing the data collection, in particular Prof. Noel Perkins, Mr. Steven Davidson, Dr. Stephen Cain, Dr. Lauro Ojeda, Ms. Rachel Vitali, Mr. Jon Mendicelli, Mr. Nathan Kossey, and Mr. Cody McKay.

References

1. Sheppard, J.M., Young, W.B.: Agility literature review: classifications, training and testing. J. Sports Sci. **9**, 919–932 (2006)
2. Spiteri, T., Newton, R.U., Nimphius, S.: Neuromuscular strategies contributing to faster multidirectional agility performance. J. Electromyogr. Kinesiol. **25**(4), 629–636 (2015)
3. Davis, D.S., Barnette, B.J., Kiger, J.T., Mirasola, J.J., Young, S.M.: Physical characteristics that predict functional performance in division I college football players. J. Strength Conditioning Res. **18**, 115–120 (2004)
4. Gil, S.M., Gil, J., Ruiz, F., Irazusta, A., Irazusta, J.: Physiological and anthropometric characteristics of young soccer players according to their playing position: relevance for the selection process. J. Strength Conditioning Res. **21**, 438–445 (2007)
5. Pauole, K., Madole, K., Garhammer, J., Lacourse, M., Rozenek, R.: Reliability and validity of the T-test as a measure of agility, leg power, and leg speed in college-aged men and women. J. Strength Conditioning Res. **14**(4), 443–450 (2000)

6. Hachana, Y., Chaabène, H., Nabli, M.A., Attia, A., Moualhi, J., Farhat, N., Elloumi, M.: Test-retest reliability, criterion-related validity, and minimal detectable change of the Illinois agility test in male team sport athletes. J. Strength Conditioning Res. **27**(10), 2752–2759 (2013)
7. Tanner, R., Gore, C.: Physiological Tests for Elite Athletes. Human Kinetics, IL (2013)
8. Spasic, M., Krolo, A., Zenic, N., Delextrat, A., Sekulic, D.: Reactive agility performance in handball; development and evaluation of a sport-specific measurement protocol. J. Sports Sci. Med. **14**, 501–506 (2015)
9. Farrow, D., Young, W., Bruce, L.: The development of a test of reactive agility for netball: a new methodology. J. Sci. Med. Sport **8**(1), 52–60 (2005)
10. Sekulic, D., Krolo, A., Spasic, M., Uljevic, O., Peric, M.: The development of a new stop'n'go reactive-agility test. J. Strength Conditioning Res. **28**(11), 3306–3312 (2014)
11. Tsitskaris, G., Theoharopoulos, A., Garefis, A.: Speed, speed dribble and agility of male basketball players playing in different positions. J. Hum. Mov. **45**(1), 21–30 (2003)
12. Sekulic, D., Spasic, M., Mirkov, D., Cavar, M.: Gender-specific influences of balance, speed, and power on agility performance. J. Strength Conditioning Res. **27**(3), 802–811 (2013)
13. Marković, G., Sekulić, D., Marković, M.: Is agility related to strength qualities?-Analysis in latent space. Collegium Antropologicum **31**(3), 787–793 (2007)
14. Yanci, J., Los Arcos, A.A., Mendiguchia, J., Brughelli, M.: Relationships between sprinting, agility one-and two-leg vertical and horizontal jump in soccer players. Kineziologija **46**(2), 194–201 (2014)
15. Sasaki, S., Nagano, Y., Kaneko, S., Sakurai, T., Fukubayashi, T.: The relationship between performance and trunk movement during change of direction. J. Sports Sci. Med. **10**(1), 112 (2011)
16. Marshall, B.M., Franklyn-Miller, A.D., King, E.A., Moran, K.A., Strike, S.C., Falvey, É.C.: Biomechanical factors associated with time to complete a change of direction cutting maneuver. J. Strength Conditioning Res. **28**(10), 2845–2851 (2014)
17. King, L.A., Horak, F.B.: Delaying mobility disability in people with Parkinson disease using a sensorimotor agility exercise program. Phys. Ther. **89**(4), 384 (2009)
18. Raya, M.A., Gailey, R.S., Gaunaurd, I.A., Jayne, D.M., Campbell, S.M., Gagne, E., Tucker, C.: Comparison of three agility tests with male servicemembers: Edgren side step test, T-test, and Illinois agility test. J. Rehabil. Res. Dev. **50**(7), 951–960 (2013)
19. Maio, G.R., Roese, N.J., Seligman, C., Katz, A.: Rankings, ratings, and the measurement of values: evidence for the superior validity of ratings. Basic Appl. Soc. Psychol. **18**(2), 171–181 (1996)
20. Sullivan, G.M., Artino Jr., A.R.: Analyzing and interpreting data from Likert-type scales. J. Grad. Med. Educ. **5**(4), 541–542 (2013)

Analysis of Pitching Skills of Major League Baseball Players

Michiko Miyamoto(✉) and Akihiro Ito

Department of Management Science and Engineering,
Akita Prefectural University, 84-4 Aza Ebinokuchi Tsuchiya,
Yurihonjo 015-0055, Japan
{miyamoto,B16D003}@akita-pu.ac.jp

Abstract. This study examines the pitcher's deciding ball after pushing a batter with two strikes of an aged pitcher group (31 to 43 year-old) and a younger pitcher group (20 to 30 year-old) by using an actual tracking data of the Major League Baseball in 2015. The regression analyses are conducted for all pitchers and for each age group on different pitch types; *i.e.*, two-seam, cutter, splitter, forkball, straight and so on. We also analyze relationships between pitchers' knocking out batters and their pitching characteristics measured by pitch movements by using a framework and empirical analyses. The results of the research model using Structural Equation Modeling show what makes the pitcher advantageous over the batter.

Keywords: Major league baseball · Tracking data · Regression analyses · Structural equation modeling

1 Introduction

Baseball games always focus on players' actual strengths and accomplishments. Winning a game is certainly related to a player's individual performance. The performance of the pitcher, particularly, becomes crucial to the outcome of a game [1–4], and pitcher performance projection is a fundamental area in baseball analysis [5]. Karakolis [6] states that pitchers are evaluated by their abilities, performances and contributions.

Best players in American (AL) and National League (NL) of Major Baseball League are young. For example, Bryce Harper of NL is twenty-three-year-old, becoming the third-youngest player to win the Baseball Writers' Association of America's National League Most Valuable Player Award in 2015, and is expected to be like Frank Robinson, who became the only player to win league MVP honors in both NL and AL, as well as winning the Triple Crown, leading the league in batting average, home runs and runs batted in. Keating [7] studied the past three decades of elite seasons by players, and found that the proportion of elite seasons by position players ages 25 and under declined sharply, beginning in the early 1990s and bottoming out at 5.9% in 2002, then it started to rise, and it has jumped sharply in 2014 and 2015, hitting a huge 34.4% in 2015.

T. Ahram (ed.), *Advances in Human Factors in Sports, Injury Prevention and Outdoor Recreation*, Advances in Intelligent Systems and Computing 603,
DOI 10.1007/978-3-319-60822-8_2

While younger players have a significant impact on the Major League scene, many 35-year-old-plus players are still contributing at a high level.

In 2007, Julio Franco played after his 49th birthday, also Roger Clemens (45), Jamie Moyer (44) and David Wells (44) are included. Barry Bonds set the record at the age of thirty-six, thirty-seven, and thirty-nine. Table 1 depicts Major League Baseball clubs by average player age in 2016. The Boston Red Sox had the roster with the highest average player age of 31 years in 2014 [8].

Table 1. Major league baseball rosters by average player age in 2016

Clubs	Ages	Clubs	Ages
Seattle Mariners	30.1	Chicago White Sox	28.7
Washington Nationals	29.8	Los Angeles Angels	28.7
Pittsburgh Pirates	29.7	New York Mets	28.6
Toronto Blue Jays	29.6	Los Angeles Dodgers	28.6
San Francisco Giants	29.5	Texas Rangers	28.5
Kansas City Royals	29.4	Houston Astros	28.4
Detroit Tigers	29.3	Cincinnati Reds	28.3
Oakland Athletics	29.3	Colorado Rockies	28.2
Atlanta Braves	29.2	Baltimore Orioles	28.1
Boston Red Sox	29	Minnesota Twins	28.1
New York Yankees	29	St. Louis Cardinals	28
Miami Marlins	29	Milwaukee Brewers	27.9
San Diego Padres	28.9	Philadelphia Phillies	27.8
Chicago Cubs	28.8	Tampa Bay Rays	27.8
Cleveland Indians	28.8	Arizona Diamondbacks	26.9

(The authors created the table based on data from Statistia [8])

Gibbs et al. [9] find that relatively older players outperform relatively younger players for the average Canadian NHL player for a period of 2000 and 2009; however, the relative age effect reversal happened among ALL-Star (2007–2009) and Olympic (1998–2010) team rosters, *i.e.*, younger players outperform older players.

Although MLB players decline after their peak in the late 20's due to declining health or skills [10], some play into their 40s.

In order to contribute to their teams and stay competitive in the MLB, the older players of MLB should perform differently from the younger players because of deteriorating their physical condition. Their performances need to be considered their physical strengths and experiences.

We have two purposes for this study. The first one is to examine the pitcher's deciding ball after pushing a batter with two strikes of an aged pitcher group (31 to 43 year-old) and a younger pitcher group (20 to 30 year-old) by using an actual tracking data of the Major League Baseball in 2015. Another purpose of this study is to investigate relationships between pitchers' striking out batters and their pitching characteristics measured by pitch movements. We use the data from the PITCHf/x®, whose service tracks and digitally records the full trajectory of live baseball pitches.

PITCHf/x® is a pitch tracking system, created by Sportvision, and has been installed in every MLB stadium since around 2006. The data includes pitch type, speed, and movement information. Pitch types are defined by mathematical models that are built around velocity, spin, and movement. It is a constantly evolving, sophisticated system.

2 Literature Review

Using longitudinal data from 86 seasons of Major League Baseball, Bradbury [10] indicates that hitters and pitchers peak around the age of 29 – later than previous estimates of 27. Fair [11] created a model which looked at peak age and how performance deviates from this high point by age. His most intriguing result was that, of players who performed a standard deviation above their expected level of performance for four seasons after the age of 28 (peak age of the study), 14 of the 17 examples played all of these seasons. Sommer [12] attempts to find the number of seasons of major league experience it takes for a player to reach his peak, by examining 5 different seasons over the past fifty years (1966, 1976, 1986, 1996, and 2006) to see how this has changed over time. A ballplayer's batting average in year t for each of his n years in the majors with a minimum of 100 at bats per season was regressed against career year [12, 13]. Sommer [12] found that the profiles have changed dramatically since the 1960s, with conceivable stronger ballplayers reaching a higher peak several years after the batting average reached a peak for regulars in 1966.

Some studies use a statistic called WAR (Wins above Replacement) as a proxy for a player's performance. WAR compares the number of wins that a player adds to his team over a replacement level player at the same position [14–18], which is an attempt by the sabermetric baseball community to summarize a player's total contributions to their team in one statistic. Furnald [19] first used WAR to examine the impact of aging in baseball, concluding that it is important for the Major League Baseball team's management to properly identify how aging is currently affecting players as well as how aging impacts players at different positions. Whiteside, et al. [20] grouped pitch types into three distinct categories: hard pitches (*i.e.*, fastball, sinkers, and cutters), breaking pitches (*i.e.*, sliders, curveballs, and screwballs), and off-speed pitches (*i.e.*, changeups, splitters, and slow curveballs), and found that the proportion of hard pitches thrown decreased significantly until the 7th inning compared with the 1st inning, while the proportions of breaking and off-speed pitches increased. Significant decreases in pitch speed, increases in vertical movement, and decreases in release height emerged no later than the 5th inning, and the largest differences in all variables were generally recorded between the 1st inning and the late innings (7–9). Pitchers were most effective during the 2nd inning and significantly worse in innings 4 and 6.

3 Research Model and Hypotheses

PITCHf/x data include the three-dimensional spatial coordinates of the ball's trajectory, along with several other pitch characteristics. Pitch speed was the exit speed of the ball from the hand. Release location and movement values were reported relative to the right-handed reference frame originating at home plate (y-axis pointing to pitching mound, z-axis pointing up, x-axis orthogonal). Horizontal release and movement values were inverted for left-handed pitchers to permit statistical analyses and interpretation (all values pertain to a right-handed pitcher). Vertical and horizontal release locations were the z and x displacements of the ball, respectively, when it left the pitcher's hand. Vertical and horizontal ball movements were the z and x displacements of the ball between the time it left the pitcher's hand and the time it crossed home plate. Zone percentage represented the percentage of pitches that were thrown in the strike zone. Each of these parameters was recorded using the PITCHf/x ball-tracking system, provided by Sportvision, Chicago, IL, which is installed in all 30 MLB stadiums [21]. PITCHf/x system has home plate as its point of origin, $\hat{y}$ points towards the pitcher, $\hat{z}$ points vertically upward, and $\hat{x} = \hat{y} \times \hat{z}$ (*i.e.*, the x axis points to the catcher's right) [22].

This paper empirically investigates factors affecting pitchers' striking out batters (hereafter, we define it as "close"), *i.e.*, "strike out," as well as "set out." "The set out" includes Called Strike, Swinging Strike, Swinging Strike - Blocked, Swinging on Pitchout, Foul Tip, Foul Tip on Bunt, Automatic Strike, Hit Into Play, Missed Bunt Attempt and Pitchout (For detailed descriptions of variables, see Table 4). Those factors are considered to affect "close" are number of pitches (pitch per at bat and balls), and amount of change in balls (pitch deflection break and pitch arc break) as shown in Fig. 3.

3.1 The Regression Models

First, we perform regression analyses to see which pitch types are closely associated to strike out and set out for the young group and the aged group as shown in model (1), and then, we conduct the structure equation modeling based on four hypotheses. Pitch types are listed in Table 2.

$$\begin{aligned} y_i = \alpha + \beta_1 FF + \beta_2 SL + \beta_3 CU + \beta_4 CH + \beta_5 FA + \beta_6 FC + \beta_7 FO + \beta_8 FS \\ + \beta_9 FT + \beta_{10} KC + \beta_{11} KN + \beta_{12} EP + \beta_{13} SI + \varepsilon \end{aligned} \tag{1}$$

where y_i : *strike out or set out.*
β = weight of each attribute and
ε = residuals.

3.2 The Structural Models

We perform the structure equation modeling to see what is affecting to "close", *i.e.*, pitchers' striking out batters. Since for a left-handed pitcher, everything goes in the opposite direction from a right-handed pitcher, we use absolute values for the analysis. The research model for the structure equation modeling will be as Fig. 1.

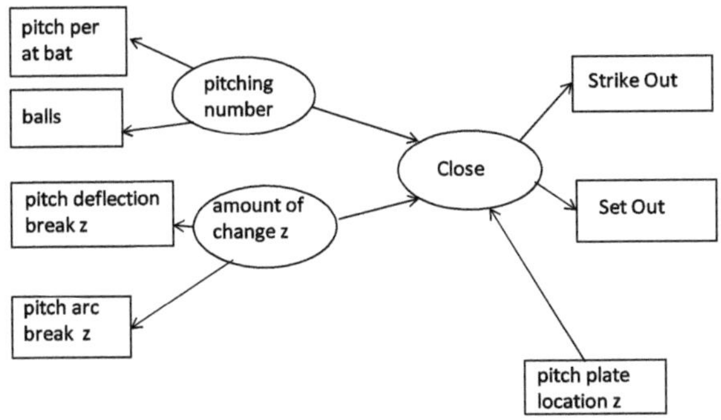

Fig. 1. The research model

More specifically, we will investigate the following three hypotheses regarding factors affecting close;

- H1: Number of pitching will affect close
- H2: Amount of change in z will affect close
- H3: Pitch plate location z will affect close

For estimating a fit between factors, advanced quantitative techniques of structural equation modeling (SEM) [23] have been employed. SEM has been established as an analytical tool, leading to hundreds of published applications per year. Overviews of the state of the method can be found in Cudeck et al. [24], Jöreskog [25], Mueller [26], and Yuan and Bentler [27]. Based on these results of analyses, we will measure how such factors, *i.e.*, shoot chance, cross front goals, players' skills, and in origination area, affect shooting.

In structural equation modeling, we consider the causalities among all variables, especially between the result and the latent variables. A latent variable enables us to find many compiled observed variables at the same time based on the notion of structure. This works for generating and verifying hypotheses to find factors and causalities.

4 Data

Pitching parameters (*i.e.*, pitch type, pitch speed, horizontal release location, vertical release location, horizontal movement, vertical movement, and percentage of pitches in the strike zone) were obtained directly from the PITCHf/x database that is made available by Data Studio Japan Inc., Japan's leading sports information provider. Each pitch is classified into 13 types: four-seam fastball, two-seam, sinker, cut ball, slider, curveball, screwball, knuckle, knuckle curve, change up, splitter, or Eephus pitch (*i.e.*, slow ball). A list of pitch types is shown in Table 2.

Descriptive statistics of variables for Pitchf/x 2015 data are shown in Table 3. An average age for this sample is 29.58 year-old. The youngest pitcher is 21, and the oldest is 43. A list of variables is shown in Table 4. Table 5 contains the Pearson correlation

Table 2. A list of pitch types

Variables	Pitch types
FF	Four-seam fastball
SL	Slider
CU	Curve
CH	Change up
FA	Straight
FC	Cut ball
FO	Fork
FS	Splitter
FT	Two-seam fastball
KC	Knuckle curve
KN	Knuckle
EP	Eephus Pitch
SI	Sinker

Table 3. Descriptive statistics

	N	Min.	Max.	Mean	Deviation
Set_out	87,048	0	1	0.74	0.438
Strikeout	87,048	0	1	0.41	0.491
FF	87,048	0	1	0.32	0.467
SL	87,048	0	1	0.19	0.396
CU	87,048	0	1	0.10	0.298
CH	87,048	0	1	0.11	0.315
FA	87,048	0	1	0.00	0.062
FC	87,048	0	1	0.05	0.219
FO	87,048	0	1	0.00	0.027
FS	87,048	0	1	0.02	0.146
FT	87,048	0	1	0.11	0.311
EP	87,048	0	1	0.00	0.020
SI	87,048	0	1	0.06	0.234
pitch_arc_break_xx	87,048	0.00	8.66	2.1442	1.25081
pitch_per_atbat	87,048	3	15	5.09	1.495
balls	87,048	0	3	1.59	1.049
pitch_plate_location_x	87,048	−4.70	4.23	−0.0184	0.75907
pitch_plate_location_ z	87,048	−2.04	6.88	2.2040	0.85922
age	87,048	21	43	29.58	3.821
		ALL	20–30	31–43(yr old))	
Number of pitches		716	482	234	

Table 4. A list of variables

Variables		Descriptions
pitch_plate_location_z		The height of the pitching position from the ground when a ball reaches the homebase
pitch_deflection_break_x		An estimated change amount in the horizontal direction; measuring the change caused by a ball rotation
pitch_arc_break_z		An estimated change amount in the vertical direction; measuring the change caused by a ball rotation
pitch_per_atbat		Number of throws in the bat
balls		Number of balls during pitching
set out	C	Called Strike
	S	Swinging Strike
	W	Swinging Strike - Blocked
	Q	Swinging on Pitchout
	T	Foul Tip
	O	Foul Tip on Bunt
	A	Automatic Strike
	X	Hit Into Play - Out(s)
	M	Missed Bunt Attempt
	Y	Pitchout - Out(s)
strike out	event_code	Batting result: if event_code = 1, then strike out = 1; else, strike out = 0

Table 5. Correlation matrix

	Set_out	Strike_out	FF	SL	CU	CH	FA	FC	FO	FS
Set_out	1	.487**	−.036**	.040**	.038**	.009**	0.006	−0.005	−0.002	.016**
Strikeout	.487**	1	−.035**	.068**	.052**	−0.004	.008*	−.010**	0.002	.008*
FF	−.036**	−.035**	1	−.338**	−.228**	−.244**	−.043**	−.159**	−.019**	−.103**
SL	.040**	.068**	−.338**	1	−.162**	−.174**	−.030**	−.113**	−.013**	−.074**
CU	.038**	.052**	−.228**	−.162**	1	−.117**	−.021**	−.076**	−.009**	−.049**
CH	.009**	-0.004	-.244**	-.174**	-.117**	1	-.022**	-.082**	-.010**	-.053**
FA	0.006	.008*	−.043**	−.030**	−.021**	−.022**	1	−.014**	−0.002	−.009**
FC	−0.005	−.010**	−.159**	−.113**	−.076**	−.082**	−.014**	1	−0.006	−.034**
FO	−0.002	0.002	−.019**	−.013**	−.009**	−.010**	−0.002	−0.006	1	−0.004
FS	.016**	.008*	−.103**	−.074**	−.049**	−.053**	−.009**	−.034**	−0.004	1
FT	−.036**	−.061**	−.240**	−.171**	−.115”	−.123**	−.022**	−.080**	−.009**	−.052**
KC	.019**	.032**	−.115**	−.082**	−.055**	−.059**	−.010**	−.039**	−0.005	−.025**
EP	−.008*	−.011**	−.014**	−.010**	−0.007	−.007*	−0.001	−0.005	−0.001	−0.003
SI	−.026**	−.043**	−.171**	−.122**	−.082**	−.088**	−.015**	−.057**	−.007*	−.037**
pitch_arc_break_xx	−.007*	−.019**	−.219**	−.269**	.075**	.223**	.040**	−.215**	.007*	.029**
pitch_per atbat	−.185**	−.161**	.059**	−.058**	−.091**	.023**	.009**	.013**	−0.004	−.009**
balls	−.211**	−.167**	.073**	−.069**	−.109**	.017**	0.001	.019**	−0.002	−.013**
pitch_plate_location_z	−.044**	−.123**	.350**	−.200**	−.174**	−.145**	−.029**	.028**	−.015**	−.082**
pitch_plate_location_x	.024**	.043**	−.038**	.101**	.022**	−.042**	−.024**	.032**	−.022**	−.043**

	FT	KC	EP	SI	pitch_arc_break_xx	pitch_per_ atbat	balls	pitch_plat_e_location	pitch_plat_e_location
Set_out	−.036**	.019**	−.008*	−.026**	−.007*	−.185**	−.211**	−.044**	.024**
Strikeout	−.061**	.032**	−.011**	−.043**	−.019**	−.161**	−.167**	−.123**	.043**
FF	−.240**	−.115”	−.014**	−.171**	−.219**	.059**	.073**	.350**	−.038**
SL	−.171**	−.082**	−.010**	−.122**	−.269**	−.058**	−.069**	−.200**	.101**
CU	−.115”	−.055**	−0.007	−.082**	.075**	−.091**	−.109**	−.174**	.022**
CH	−.123**	−.059**	−.007*	−.088**	.223**	.023**	.017**	−.145**	−.042**

(*continued*)

Table 5. *(continued)*

	FT	KC	EP	SI	pitch_arc_break_xx	pitch_per_ atbat	balls	pitch_plat_e_location	pitch_plat_e_location
FA	−.022**	−.010**	−0.001	−.015**	.040**	.009**	0.001	−.029**	−.024**
FC	−.080**	−.039**	−0.005	−.057**	−.215**	.013**	.019**	.028**	.032**
FO	−.009**	−0.005	−0.001	−.007*	.007*	−0.004	−0.002	−.015**	−.022**
FS	−.052**	−.025**	−0.003	−.037**	.029**	−.009**	−.013**	−.082**	−.043**
FT	1	−.058**	−.007*	−.087**	.325**	.044**	.054**	.082**	−.040**
KC	−.058**	1	−0.003	−.042**	.024**	−.048**	−.052**	−.101**	.022**
EP	−.007*	−0.003	1	−0.005	.009**	−0.003	−.007*	−.007*	0.005
SI	−.087**	−.042**	−0.005	1	.227**	.035**	.045**	.047**	−.023**
pitch_arc_break_xx	.325**	.024**	.009**	.227**	1	.012**	.019**	−.059**	−.015**
pitch_per atbat	.044**	−.048**	−0.003	.035**	.012**	1	.822**	.029**	−.033**
balls	.054**	−.052**	−.007*	.045**	.019**	.822**	1	.029**	−.034**
pitch_plate_location_z	.082**	−.101**	−.007*	.047**	−.059**	.029**	.029**	1	−.104**
pitch_plate_location_x	−.040**	.022**	0.005	−.023**	−.015**	−.033**	−.034**	−.104**	1

** The correlation coefficient is significant (two sides) at the 1% level. * at the 5% level.

coefficient between all pairs of twenty-one variables with the two-tailed significance of these coefficients. All variables correlate fairly well and are statistically significant, and none of the correlation coefficients are particularly large; therefore, multicollinearity is not a problem for this data.

5 Results of Analyses

5.1 The Regression Models

We set "strike out," "set out" or "being struck (including four balls)" as an event. The missing values were excluded. Data are limited only when the events occurred and when the pitchers push a batter with two strikes. A regression analysis is performed for each pitch type.

As for the independent variable, a dummy variable is created for each pitch type. The target variable is set to "1" for strike out or set out, and "0" for otherwise. In other words, either the strike out or set out indicate whether the pitcher struck the batter in any way. The results of the regression on the 13 different pitch types are summarized in Table 6 (dependent variable: set out) and Table 7 (dependent variable: strike out).

The results for "set out" shows that all ball types are positive and statistically significant. For all and both age groups, the four-seam fast ball has the highest coefficient. Those which the younger group has the higher coefficient than the older group are Four-seam, Slider, Curve, Straight, while the older group has the higher coefficient in Change up, Cut ball, Two-seam fastball, Knuckle curve, Knuckle, and Sinker.

Table 6. The result of regression analysis (dependent variable: set out)

		Overall			20–30 years old			31–43 years old		
		Coef	t.stat	p.value	Coef	t.stat	p.value	Coef	t.stat	p.value
FF	Four-seam fastball	0.473	275.741	.000	0.483	225.594	0.000	0.456	158.593	0.000
SL	Slider	0.398	231.735	.000	0.407	190.348	0.000	0.380	132.169	0.000
CU	Curve	0.289	168.131	.000	0.299	139.712	0.000	0.269	93.525	0.000
CH	Change up	0.292	170.046	.000	0.287	134.098	0.000	0.301	104.597	0.000
FA	Straight	0.057	32.985	.000	0.067	31.242	0.000	0.031	10.603	0.000
FC	Cut ball	0.191	111.118	.000	0.169	78.930	0.000	0.225	78.278	0.000
FO	Fork	0.022	12.893	.000	0.028	12.884	0.000	–	–	–
FS	Splitter	0.135	78.888	.000	0.110	51.321	0.000	0.173	59.980	0.000
FT	Two-seam fastball	0.266	155.035	.000	0.264	123.390	0.000	0.270	93.888	0.000
KC	Knuckle curve	0.152	88.335	.000	0.149	69.738	0.000	0.156	54.235	0.000
KN	Knuckle	0.045	26.352	.000	–	–	–	0.076	26.391	0.000
EP	Eephus Pitch	0.013	7.749	.000	0.013	5.860	0.000	0.015	5.093	0.000
SI	Sinker	0.195	113.489	.000	0.186	86.919	0.000	0.210	73.052	0.000
R		0.862			0.862[a]			0.863[a]		
R Square[b]		0.743			0.743			0.745		
Adjusted R Square		0.743			0.743			0.745		
Std. Error of the Estimate		0.436			0.437			0.436		

[a] Dependent Variable: set out. [b] Linear Regression through the Origin

Table 7. The result of regression analysis (dependent variable: strike out)

		Overall			20–30 years old			31–43 years old		
		Coef	t.stat	p.value	Coef	t.stat	p.value	Coef	t.stat	p.value
FF	Four-seam fastball	0.339	130.524	0.000	0.342	106.251	0.000	0.333	75.891	0.000
SL	Slider	0.328	126.223	0.000	0.336	104.301	0.000	0.312	71.125	0.000
CU	Curve	0.238	91.669	0.000	0.251	77.998	0.000	0.212	48.291	0.000
CH	Change up	0.210	80.885	0.000	0.209	64.830	0.000	0.213	48.424	0.000
FA	Straight	0.046	17.616	0.000	0.053	16.366	0.000	0.030	6.728	0.000
FC	Cut ball	0.135	52.137	0.000	0.117	36.350	0.000	0.165	37.465	0.000
FO	Fork	0.019	7.177	0.000	0.023	7.169	0.000	–	–	–
FS	Splitter	0.100	38.693	0.000	0.091	28.182	0.000	0.118	26.927	0.000
FT	Two-seam fastball	0.165	63.713	0.000	0.165	51.305	0.000	0.166	37.817	0.000
KC	Knuckle curve	0.129	49.820	0.000	0.125	38.852	0.000	0.137	31.224	0.000
KN	Knuckle	0.026	10.121	0.000	–	–	–	0.045	10.150	0.000
EP	Eephus Pitch	0.004	1.733	0.083	0.004	1.310	0.190	0.005	1.141	0.254
SI	Sinker	0.121	46.786	0.000	0.118	36.775	0.000	0.128	29.061	0.000
R		0.643			0.647[a]			0.637[a]		
R Square[b]		0.413			0.418			0.405		
Adjusted R Square		0.413			0.418			0.405		
Std. Error of the Estimate		0.488			0.488			0.486		

[a] Dependent Variable: strike out. [b] Linear Regression through the Origin

A two seam fastball, much like a sinker or cutter (cut fastball), is gripped slightly tighter and deeper in the throwing-hand than the four-seam fastball. This pitch generally is thought of as a "movement pitch," as opposed to the four-seam fastball, which is primarily thought of as a "straight pitch" [28]. The results imply that younger pitchers are throwing more straight pitches, while the older pitchers are throwing more movement pitches.

The results for "strike out" show that all ball types, except Eephus Pitch, are positive and statistically significant. Eephus Pitch is positive and statistically significant at a 10% level for an overall result, but positive and not significant for both age groups.

A coefficient for the Eephus Pitch is very small, as well. Those of the younger group that has the higher coefficient than the older group are Four-seam, Slider, Curve, Straight, while the older group has the higher coefficient in Change up, Cut ball, Knuckle curve, Knuckle, and Sinker. There is almost the same level of coefficient in Two-seam fastball for both groups. The results for strike out also imply that younger pitchers are throwing more straight pitches, while the older pitchers are throwing more movement pitches.

5.2 The Structural Equation Models – Results of Hypotheses

Based on the research model depicted in Fig. 1, we test the efficacy of the structural equation model that was conducted by AMOS 24. Among different pitch types, we select the four-seam and the change up, which had higher coefficients on the regression analyses for the young and the aged group respectively. The major results of analysis for the four seam ball for the young group is shown Fig. 2, and those for the aged group is shown in Fig. 3, respectively. The results for the four seam ball for the young group and those for the aged group, and those for the change up for the young and the aged group are shown in Table 8.

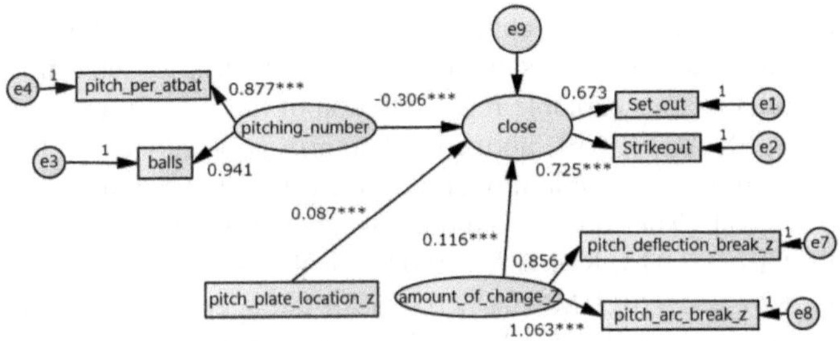

Fig. 2. Four seams (young)

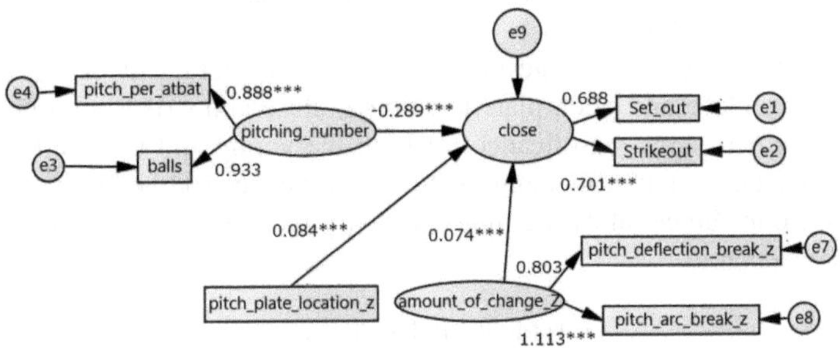

Fig. 3. Four seams (aged)

Table 8. The path coefficients of research models (standard weights)

Construct		Young	Aged	Young	Aged
		Four-seam first ball (FF)		Change up (CH)	
close	← pitching_number	−0.306***	−0.289***	−0.169***	−0.068***
close	← amount_of_change_z	0.116***	0.074***	−0.026	−0.012
close	← pitch_plate_location_z	0.087***	0.084***	−0.233***	−0.229***
Set out	← close	0.673	0.688	0.557	0.419
Strikeout	← close	0.725***	0.701***	0.852***	1.082***
balls	← pitching_number	0.941	0.933	0.887	0.744
pitch_per_atbat	← pitching_number	0.877***	0.888***	0.887***	1.051***
pitch_deflection_break z	← amount_of_change_z	0.856	0.803	0.951	1.334
pitch_arc_break_z	← amount_of_change_z	1.063***	1.113***	0.847***	0.631

*** Denotes significance at 1%

The path diagram highlights the structural relationships. In these diagrams, the measured variables are enclosed in boxes, latent variables are circled, and arrows connecting two variables represent relations, and open arrows represent errors.

When SEM is used to verify a theoretical model, a better goodness of fit is required for SEM analysis; the better the fit, the closer the model matrix and the sample matrix. By means of various goodness-of-fit indexes, including the Goodness-of-Fit statistic (GFI) and the adjusted goodness-of fit (AGFI) [29], the comparative fit index (CFI) [30], and the root mean squared error of approximation (RMSEA) [31], estimated matrix can be evaluated against the observed sample covariance matrix to determine whether the hypothesized model is an acceptable representation of the data. In general, incremental fit indexes (*i.e.*, GFI, AGFI, CFI) above 0.90 signify good model fit. RMSEA values lower than 0.08 signify acceptable model fit, with values lower than 0.05 indicative of good model fit [31]. The research model is shown in Table 8 as GFI = 0.975, AGFI = 0.942, CFI = 0.973, RMSEA = 0.049 for the young group and GFI = 0.979, AGFI = 0.951, CFI = 0.976, RMSEA = 0.045 for the aged group (see Table 9).

The path coefficient for structural models of the four-seam first ball suggested that the regression coefficient between close and pitching number; close and amount of change z; close and pitch plate location z, close and strike out; pitch per at bat and pithing number; pitch arc break z and amount of change z show significance for both the young and the aged group.

Those of the change up suggested the regression coefficient between close and pitching number; close and pitch plate location z, close and strike out; pitch per at bat and pithing number; show significance for both the young and the aged group, while pitch arc break z and amount of change z show significance for the young group but not for the aged group. Since all of the indexes satisfy the cut-off values, these results are regarded as acceptable.

Table 9. Reliability tests

FIT indices	Recommended level	Young	Aged
CMIN/DF	5.0 (Wheaton et al. [32]) ~ 2.0 (Tabachnick and Fidell [29])	89.925	37.995
GFI	>0.90 (Tabachnick and Fidell [29])	0.975	0.979
AGFI	>0.90 (Tabachnick and Fidell [29])	0.942	0.951
CFI	>0.90 (Bentler [30])	0.973	0.976
RMSEA	<0.08 (Browne and Cudeck [31])	0.049	0.045
AIC	Smaller values suggest a good fitting (Akaike [33])	3333.309	1463.827
p-value	>0.05	0.000	0.000

The results of the research models for the young group and the old group for the four-seam first ball show the following three findings;

- H1: Number of pitching is significantly, negatively affecting close
- H2: Amount of change in z is significantly, positively affecting close
- H3: Pitch plate location z is significantly, positively affecting affect close

And those for the change-up show the following three findings;

- H1: Number of pitching is significantly, negatively affecting close
- H2: Amount of change in z is significantly, negatively affecting close
- H3: Pitch plate location z is significantly, positively affecting affect close for the young group, but not statistically significant for the aged group.

The results of the structure models imply that there is not so much of a difference between two age groups in terms of factors relating to the close.

6 Conclusion and Future Study

We conducted two different analyses, *i.e.*, the regression analyses and the structural equation models, in this study. The results from the regression analyses imply that younger pitchers are throwing more straight pitches, while the aged pitchers are throwing more movement pitches. The results of the structure models imply that there is not so much difference between the two age groups in terms of how pitchers' striking out batters.

Movement refers to the spin-induced deflection, and break refers to the maximum bend in the pitch. The arc of a curveball bends much more than a fastball [34]. Older pitchers differ from their younger counterparts in a variety of physical and mental dimensions. Older pitchers may lose their physical strength somewhat, while they have gained their skills in pitching through their careers. We did not study the data in terms of differences in left-arms and right-arms, and that will be our future study.

References

1. Kindall, J.: Sports Illustrated Baseball: Play the Winning Way. Time, Inc., Lanham, MD (1993)
2. Lopez, A., Kirkgard, J.: Coaching Baseball Successfully. Human Kinetics, Champaign, IL (1996)
3. Yeh, C.S., Lee, M.S., Zhu, F.L.: Baseball Bible. Business Weekly Publication, Inc, Taipei (2003)
4. Chiu, S., Hsiao, C., Wu, H.: Measuring pitchers' performance using data envelopment analysis with advanced statistics. Contemp. Manage. Res. **11**(4), 351–384 (2015). doi:10.7903/cmr.14157
5. Shu, P.E.S.: Arsenal/zone rating: a PitchF/X based pitcher projection system. In: 2016 MIT Sloan Sports Analytics Conference Research Paper (2016)
6. Karakolis, T.: Measuring a Pitcher's Ability, Performance, and Contribution, Fan Graphs Community Research. http://www.fangraphs.com/community/measuring-a-pitchers-ability-performance-and-contribution/ (2013). Accessed 16 Feb 2017
7. Keating, P.: Better, Faster, Younger: Why Baseball's Young Stars are Its Best in 20 years, ESPN The Magazine. http://www.espn.com/mlb/story/_/id/14946288/better-faster-younger-mlb-best-stars-20-years (2016). Accessed 17 Feb 2017

8. Statistia: Major League Baseball Rosters by Average player age in 2016. https://www.statista.com/statistics/236223/major-league-baseball-clubs-by-average-age-of-players/ (2016). Accessed 17 Feb 2017
9. Gibbs, B.G., Jarvis, J.A., Dufur, M.J.: The rise of the underdog? The relative age effect reversal among Canadian-born NHL hockey players: a reply to Nolan and Howell. Int. Rev. Sociol. Sport. **47**(5), 644–649 (2012)
10. Bradbury, J.: Peak athletic performance and ageing: evidence from baseball. J. Sports Sci. **27**(6), 599–610 (2009)
11. Fair, R.C.: Estimated age effects in baseball. J. Quant. Anal. Sports **4**(1), 1 (2008). Article 1
12. Sommers, P.: The changing hitting performance in major league baseball, 1966-2006. J. Sports Econ. **9**(4), 435–440 (2008)
13. Scully, G.W.: The Business of Baseball. University of Chicago Press, Chicago (1989)
14. Gennaro, V.: Diamond Dollars: The Economics of Winning in Baseball. Maple Street Press, Hingham, MA (2007)
15. Forman, S.: Player Wins Above Replacement. http://www.baseball-reference.com/blog/archives/ 6063 (2010) Accessed 19 Feb 2017
16. Slowinski, S.: What is WAR? http://www.fangraphs.com/library/index.php/misc/war/ (2010). Accessed 17 Feb 2017
17. Bradbury, J.: Hot Stove Economics: Understanding Baseball's Second Season. Copernicus Books, New York (2011)
18. Baumer, B.S., Jensen, T.: openWAR: an open source system for evaluating overall player performance in major league baseball. J. Quant. Anal. Sports **11**(2), 69–84 (2015)
19. Furnald, N.A.: The Impact of Age on Baseball Players' Performance, Unpublished manuscript. Colgate University (2012)
20. Whiteside, D., Martini, D.N., Zernicke, R.F., Goulet, G.C.: Changes in a starting pitcher's performance characteristics. Across the duration of a major league baseball game. Int. J. Sports Physiol. Perform. **11**(2), 247–254 (2016)
21. Fast, M.: Four plus pitches. The hardball times. http://www.hardballtimes.com/main/article/four-plus-pitches/ 23 June 2010. Accessed 21 Feb 2017
22. Kogan, D.: The anatomy of a pitch: doing Physics with PITCHf/x data. Phys. Teach. **47**(7), 412–416 (2009)
23. Byrne, B.M.: Structural Equation Modeling Using AMOS. Basic Concepts, Applications, and Programming, 2nd edn. Taylor & Francis, New York (2010)
24. Cudeck, R., du Toit, S., Sörbom, D. (eds.): Structural Equation Modeling: Present and Future. Scientific Software International, Chicago (1992)
25. Jöreskog, K.G.: New developments in LISREL—analysis of ordinal variables using polychoric correlations and weighted least squares. Qual. Quant. **24**, 387–404 (1990)
26. Mueller, R.: Structural equation modeling: back to basics. Struct. Equ. Model. **4**, 353–369 (1997)
27. Yuan, K.H., Bentler, P.M.: Mean and covariance structure analysis: theoretical and practical improvements. J. Am. Stat. Assoc. **92**, 767–777 (1997)
28. Ellis, S.: Pitching Grips- How to Grip and Throw Different Baseball Pitches. http://www.thecompletepitcher.com/pitching_grips.htm (2017). Accessed 23 Feb 2017
29. Tabachnick, B.G., Fidell, L.S.: Using Multivariate Statistics, 5th edn. Allyn and Bacon, New York (2007)
30. Bentler, P.M.: Comparative fit indexes in structural models. Psychol. Bull. **107**, 238–246 (1990)
31. Browne, M.W., Cudeck, R.: Alternative ways of assessing model fit. Sociol. Methods Res. **21**, 230–239 (1993)

32. Wheaton, B., Muthen, B., Alwin, D.F., Summers, G.: Assessing reliability and stability in panel models. Sociol. Methodol. **8**(1), 84–136 (1977)
33. Akaike, H.: A New Look at the Statistical Model Identification. IEEE Trans. Autom. Control **19**(6), 716–723 (1974)
34. Fast, M.: What the heck is PITCHf/x. In: Distelheim, J., Jacobs, B. (eds.) The Hardball Times Baseball Annual 2010, pp. 153–158. ACTA Sports, Chicago (2009)

Methodology for the Assessment of Clothing and Individual Equipment (CIE)

Leif Hasselquist(✉), Marianna Eddy, K. Blake Mitchell, Clifford L. Hancock, Jay McNamara, and Christina Caruso

U.S. Army Natick Soldier Research, Development and Engineering Center, Natick, MA, USA
leif.hasselquist.civ@mail.mil

Abstract. The Army has been searching for a repeatable and sensitive way to evaluate clothing and individual equipment (CIE), including personal protective systems, for increase in mission flexibility, mobility, and protection. The goal of this methodology was to have Soldiers engage in an operationally relevant scenario and to provide a continuous, fatiguing set of tasks, mimicking movement to and action on an objective. By using an interdisciplinary approach (incorporating biomechanical, cognitive psychology, and human factors expertise), we have demonstrated a comprehensive methodology that addresses the Warfighter as a system. The paradigm of evaluating the Warfighter from a cognitive, physiologic, and performance approach allows the ability to analyze how these processes interact. This approach gives a complete picture of what the Warfighter's challenges are in complex scenarios.

Keywords: Biomechanics · Human factors · Cognitive · Physical performance · Marksmanship

1 Introduction

The Army continually seeks to improve the equipment used to protect the individual Soldier. The Army has typically assessed the acceptability of next-generation or novel protective equipment through human factors or limited user evaluations of the items by Soldiers. The results of these assessments have consisted mainly of subjective data regarding the test items, in the form of participants' comments and opinions. While these assessments have gleaned useful information, they are not comparable across different evaluations and have not investigated the quantitative effects test items have on Soldiers' performance of militarily relevant activities. While laboratory studies provide a rich literature on cognitive and physical performance under conditions of load carriage that simulate some of the mission relevant conditions Soldiers are asked to perform, there is a need to move evaluations of equipment to field relevant scenarios to ensure validity in the operational context. An assessment is required that captures objective data, collected

PAO#: U17-122

T. Ahram (ed.), *Advances in Human Factors in Sports, Injury Prevention and Outdoor Recreation*, Advances in Intelligent Systems and Computing 603,
DOI 10.1007/978-3-319-60822-8_3

over a long duration in order to characterize the effects of physical fatigue induced by Soldier equipment on cognitive and physical performance. The goal of this test methodology was to have Soldiers engage in operationally relevant tasks and to provide a continuous, fatiguing set of tasks, mimicking movement to an objective and action on objective. This methodology allows for the inclusion of a statistically relevant number of participants, with a number of test conditions, in a reasonable amount of time and in varied environments depending on the goals of the evaluation. This scenario consists of a baseline rifle marksmanship test, 3 mi road march, Load Effects Assessment Program-Army (LEAP-A) obstacles, Military Operations on Urban Terrain (MOUT) room clearing exercise, and second 3 mi road march, and finally a rifle marksmanship performance test (Fig. 1). Throughout each of these events, physical, biomechanical, and cognitive measures are recorded. Which allows for comparison of marksmanship and road march performance in a rested and fatigued state.

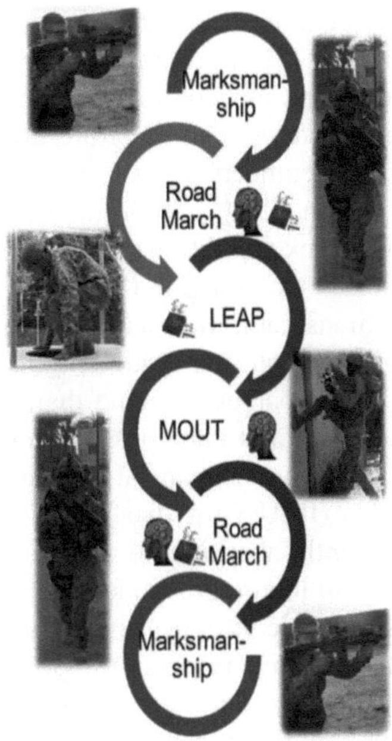

Fig. 1. Overview of test scenario

As structured within our methodology the participants complete the test scenario three times, which provides within-subjects measures of performance when donning three different equipment configurations. This test methodology is designed to have the flexibility to be used with a number of equipment conditions in varied environments and for an assortment of Soldier relevant performance tasks.

2 Background

2.1 Biomechanical and Physiologic Measures During Soldier Performance

The data acquired under the current methodology consisted of walking kinematics during the prolonged march, in addition to running, balance and change of direction related kinematics during the LEAP-A under different configurations. A large amount of research has been completed on evaluating the effects of added load on the body. Much of the research related to the carrying of loads by Soldiers has involved quantification of the effects that the weight of the external load has on energy consumption, with the rate of oxygen uptake $\dot{V}O2$ used as an indicator of energy consumption [1]. Findings from research guides the development of military load-bearing equipment, set design goals for maximum weights of Soldier loads, and inform military commanders in the field regarding procedures for conducting foot marches [2–4]. Very few of these studies include biomechanical measures in association with prolonged bouts of load carrying. Gait data were acquired in a study conducted by Frykman et al., but only at the beginning and the end of a period of load carrying not in a continuous dynamic format [5]. Furthermore, no studies sited in the literature report simultaneously recording physiological and biomechanical responses to prolonged load carrying in a field setting. The literature also suggests that, in addition to the weight it represents, CIE encumbers or restricts body movements, further contributing to negative effects on performance [6]. In terms of body kinematics, studies have shown that increases in backpack weight result in increases in the forward inclination of the trunk. Martin and Nelson examined the effects of load weight on spatial and temporal gait variables and demonstrated that, with an increase in pack load, both men and women had a higher step rate, shorter stride length, shorter swing time, and increased double support time [7]. Harman et al. found that, with an increase in load weight, a delay in foot push-off occurred and knee flexion increased [8].

The biomechanical data in the proposed methodology utilizes inertial measurement units (IMUs) for data collection in a field environment. Research organizations are currently working to systematically advance IMU technologies as they pertain to human motion, to achieve a prototype for reliably assessing warfighter performance in naturalistic settings. The University of Michigan has been involved with NSRDEC in a collaborative research program to develop the use of IMUs for field research [9]. Cain et al. have been able to calculate a wide range of metrics for gait in an outdoor, non-laboratory setting. Many of these metrics are identical to those used in a gait lab (stride length, speed, joint angles, and torso angles) [10].

2.2 Impact of Load Carrying on Physical and Task Performance

The current methodology utilizes a number of militarily relevant physical activities, including an obstacle course run, MOUT scenario, marksmanship and road marching. Studies investigating the effects of load carriage on agility and performance of Soldier-related physical tasks have, like the research pertaining to energy cost and biomechanics, been focused mainly on assessing the impact of load mass. Holewijn and Lotens reported that backpack mass contributed to performance decrements in maneuvering through obstacles, hand-grenade throwing, running, and jumping [11]. Similarly, other researchers found that performance decreased for an agility run with an increase in load and Knapik et al. found that times to complete a 20.0-km (12.4-mile) road march increased as load mass increased [12–15]. Only a limited amount of research reports the effects of different designs of load-carriage equipment have on performance. Harman et al. reported that female Soldiers wearing the All-purpose Lightweight Individual Carrying Equipment (ALICE), as compared to the Modular Lightweight Load-carrying Equipment (MOLLE), performed better in terms of obstacle course completion times and times to complete such activities as going from a standing position to a prone position and returning to a standing position [14].

Timed runs of obstacle courses, which require such activities as jumping, crawling, climbing, and balancing, have been used extensively in studies to evaluate different designs of load-carriage equipment [15, 16]. Pandorf et al. reported that time to complete an obstacle course is a highly reliable measure (intraclass correlation coefficient of .92) [17]. Scientists have used a combined obstacle and MOUT course in previous studies of load carriage equipment. Hasselquist et al. tested rucksacks varying in prototype design. They found that course completion times were sensitive to differences in weight on the body when comparing a medium rucksack to no rucksack, but not pack design [18]. LaFiandra et al. examined differences in obstacle and MOUT course times for three load-carriage systems that were approximately equal in weight and had similar center of masses (COMs). LaFiandra et al. did not obtain significant differences in completion times among the systems. The particular courses used in the current study consist of the obstacle course portion of the LEAP-A followed by an indoor MOUT course [16].

In the current study, marksmanship data collection used the FN Expert weapon simulator system. Marksmanship utilized the FN Expert simulator system during pre- and post-scenario testing and during the MOUT course. Weapon simulator systems have been successfully used in a number of research efforts to measure the effect of postural stability on marksmanship performance, as well as the effects of creatine and caffeine supplementation on marksmanship during stress induced training research [19–22]. The FN Expert system is able to record and display muzzle trace, location of miss or hit for a shot, and allows for analysis of marksmanship performance. The FN Expert software records shot performance data real-time and presents multiple statistical calculations as well as individual shot scores and time between shots.

2.3 Physiologic Dependent Measures During Physical Performance

The dependent measures recorded for the physical tests in the field environment are typically times to completion or the number of elements completed. In the current evaluation, $VO_{2\ peak}$ and heart rate reserve (HRR) also monitored. The HRR, also known as the Karoven method, is the most accurate way of establishing exertion levels in a field setting [23]. Percentages of HRR accurately reflects the same percentages of $\dot{V}O2$ reserve $\dot{V}O2R$. The American College of Sports Medicine has adopted $\dot{V}O2R$ and HRR as the primary means of establishing exercise intensity. The one to one relationship of percentage of $\dot{V}O2R$ (% $\dot{V}O2R$) and percentage HRR (% HRR) has been confirmed in a variety of studies in a wide range of populations [24–26]. In this methodology energy cost is expressed as percentage of $\dot{V}O2R$ and heart rate as percentage of HRR.

2.4 Cognitive Performance During Physical Exertion

The available literature relating physical exertion to cognition is often confusing and contradictory [27–31]. Exercise appears to facilitate performance on some tasks under some conditions, while inhibiting performance on other tasks, or on the same tasks under different conditions. Findings of facilitation, inhibition, or no effect are common, sometimes under disappointingly similar conditions. Differences in the nature, duration, and intensity of the physical exertion undergone by subjects in these studies, as well as in the type of cognitive task employed and the fitness level of the participants, may explain this complicated and contradictory picture.

In limited cases, cognitive performance has been measured during exercise, but the duration and intensity were likely insufficient to simulate the demands in naturalistic or more extreme exertion situations [28]. However, these studies that investigated cognition during exercise seemed more likely to detect changes in cognitive performance, even when the physical requirements of the exercise were relatively mild [32–34]. In addition, it is possible that the act of maintaining attention and processing information during exercise increases workload and produces differential task performance compared to when an individual is able to concentrate simply on the physical activity.

As for a mechanism underlying the patterns of performance during exercise, Dietrich offers the transient hypofrontality hypothesis as well as the more recent reticular-activating hypofrontality (RAH) model [35, 36]. According to these hypotheses/models, there is a decrease in frontal neural activity due to the demands of exercise. Specifically, because exercise recruits activity in motor pathways (e.g. primary and secondary motor cortices, basal ganglia, cerebellum) as well as sensory (e.g. primary sensory cortex) and autonomic pathways (e.g. hypothalamus), structures supporting higher-level cognitive processes, including the prefrontal cortex, are disengaged, whereas motor performance is enhanced. Alternative explanations for cognitive performance during exercise can be linked to activation of the autonomic nervous system, resulting in increased catecholamine activity, including norepinephrine and dopamine as explanations for enhanced reaction time while exercising [37].

3 Methodological Approach

The methodology described in this paper is a within-subjects design, that is, participants serve as their own control. When utilizing this methodology outside the context of this evaluation to assess the effects of CIE on the Soldier it is recommended that similar Soldier physical tasks be used and in the same order of the scenario: (1) Marksmanship task (simulation or live fire task may vary), (2) road march (terrain may vary), (3) Soldier obstacles (order and number may vary), (4) MOUT (specific course in MOUT clearing exercise may vary), (5) repeated road march and (6) repeated marksmanship task. The techniques used for analysis of IMU, HR, cognition, and marksmanship data are essential components of the test methodology detailed in this paper. The field test scenario is a series of Soldier performance events conducted for each randomized condition. Implementation of this methodology allows for comparison of numerous equipment configurations and Soldier outcomes.

In the current study, the 62 volunteer participants wore each of three conditions on different test days separated by at least one rest day. Table 1 outlines a comprehensive set of assessment tools used to address the challenges of Soldier performance research in field environments.

The specific IMUs, HR monitors, weapon simulator, and Soldier relevant obstacles detailed in this study were our preferred equipment. This methodology is not limited to the use of our specific equipment or defined tasks. Similar equipment with reliable measurement capabilities, alternate Soldier physical tasks, and different Soldier relevant obstacles may be used depending on the stated objectives of the assessment.

Table 1. Comprehensive assessment parameters

Biomechanical	Physiological	Performance	Cognitive	Human factors
Temporal Measures Kinematics PCA	Heart rate Heart rate reserve $VO_{2\ peak}$, fatigue	Reaction time Accuracy Time to complete	Auditory task Response inhibition Marksmanship Judgement Decision making	Fit/Sizing Perceived exertion Basic movements Pain, discomfort

When participants arrived for testing, they first donned the IMUs, heart rate monitor, and condition configuration for the test scenario. Each participant then completed the full scenario in approximately three hours. The participant started the scenario by first performing the pre-marksmanship, followed by a pause to synchronize the spacing of 20-minute intervals between participants as they proceeded with the 3-mile paced march. This 20-minute spacing was necessary to avoid overlap on the road march. The march was at 3-mph. and conducted in an assigned condition. Participants wore in-ear headphones attached to the tablet carried in their pack and responded using a hand-held USB response device. An assault pack was worn during the march when a condition other than a baseline no load condition was utilized and a small backpack for the tablet may be worn in the baseline no load condition and a mock M4 is carried. Upon completion of the road march, participants doffed the assault pack or tablet pack, mock weapon, headphones, and response device, and proceeded to the LEAP-A obstacle course. They completed the obstacles in the prescribed manner with a maximal effort. After completing the LEAP course, participants immediately proceeded to the MOUT received the Noptel weapon and donned a go-pro camera on their helmet. The participant then completed room clearing shoot/don't shoot task. After completing the MOUT, participants returned to the start of the road march where they donned headphones, hand-held response device, mock weapon, and either an assault pack or small backpack with the tablet (depending on condition) and executed the 3-mi road march at 3 mi/h for a second time. The march route was identical to the route undertaken earlier in the scenario. After completion of the second road march, participants doff the assault pack, headphones, mock weapon, response device, and proceed to the weapon simulator for post-marksmanship testing. Prior to starting and after completing each scenario event, participants give RPEs and Mission Performance Ratings upon completion of each event. Completion of marksmanship task marked the end of the scenario, participants then completed the pain and discomfort scale and human factors questionnaires.

4 Metrics

4.1 Physiologic and Biomechanical Metrics

The determination of $\dot{V}O2_{2\mathrm{Peak}}$ was on a day on which no other physical activities were scheduled. Peak oxygen uptake comprised of using a continuous, uphill, and stepwise treadmill protocol. Heart rate, ratings of perceived exertion, and cardiorespiratory measures from the metabolic system (oxygen uptake, ventilation, etc.) were continuously monitored during the test in order to assess physiological responses. This evaluation utilized the recording of physiological measures and the application of HRR data.

For each of the scenario components (road marches, LEAP-A, MOUT, and marksmanship) and during the entire field test, the biomechanical IMU data was collected continuously. Selection of specific segments of the scenario events to analyze were identified based on unique algorithms. The University of Michigan and NSRDEC have collaborated on the development of biomechanical IMU algorithms used in this assessment and these are essential to assess performance changes that occur due to CIE worn by the Soldier. For these analyses, linear acceleration and angular velocity data are processed using custom Matlab (Mathworks, Natick, MA) programs to calculate trunk kinematic and spatiotemporal parameters during all study tasks. Evaluation consisted of specific segments of each task within the scenario. Metrics identified from task analyses included: stride length for every stride (left and right); stride duration for every stride (left and right); average stride speed for every stride, or average speed for each stride (left and right) and percentage of variation in a the trunk and foot angular velocity during a stride that can be described by a rotation about a single body-fixed axis (for each segment for every stride). In addition to these metrics, the following were also calculated: stride width for every stride (left and right); foot yaw for every stride (left and right); torso medial-lateral lean angle versus time; and torso anterior-posterior lean angle versus time.

4.2 Cognitive Metrics

Cognitive performance was analyzed during the road march and MOUT scenario segments of this methodology. For the road march, participants donned in ear noise cancelling headphones (M4 Electronics). Additionally, a Samsung Slate 7 tablet was placed in the Army ruck for the two loaded equipment conditions and in a small backpack for the no equipment condition. The participants also held a USB response device consisting of a single button along with a slung weapon. Participants performed an auditory go/no-go task during the road march. For this task, presentation consisted of AK47 and M4 gunfire sounds through the headphones. The participants were asked to only respond to the AK47 but not M4 gunfire by pressing the response device button. This task was repeated 5 times throughout the 60-minute march with each iteration lasting 5 min followed by 7-minute breaks. A 5-minute segment of the task contained 125 trials, for a total of 625 trials for each road march. This go/no-go task had a frequent go stimulus (AK-47) that set up a pre-potent response (one that was difficult to withhold) by having a large proportion of go trials (80%) and relatively few no-go trials (20%). Therefore, participants anticipated go trials, making it difficult to inhibit responding to no-go trials. Thus, the go/no-go task measured executive control and required sustained attention and constant monitoring for the no-go stimulus. This probed response inhibition, which is a task very relevant for Soldiers in operational contexts. Previous laboratory research by Eddy et al. found decrements in go/no go task performance (a measure of executive functioning) when Soldiers carry a load compared to when they do not. An auditory version of the task was chosen given its field portability (e.g., participants responded to auditory rather than visual stimuli allowing for stimulus presentation to occur as participants moved through the march), and has been shown to be sensitive to load effects in previous studies [38].

The completion of the MOUT scenario transpired in sequence with no rest breaks and included hallways, rooms, and stairs. Cognitive performance was measured on a shoot/don't shoot decision making task while decision accuracy and reaction time measures were collected. This task mirrored the cognitive processes in the auditory go/no-go task during the road march, but because we were able to set-up a test environment (the MOUT), we chose to use a shoot/don't shoot task to make the task as operationally relevant as possible. The shoot/don't shoot task measured decision accuracy on shooting at targets placed within the MOUT environment using the FN Expert weapon simulator system. Participants were briefed on which targets were threats (targets that should be shot) and which were non-threatening (targets that should not be shot) during the orientation session and prior to completing testing. Some examples of threatening objects on the targets included: a pistol, a machine gun, and a bomb; some examples of non-threatening objects on the targets included: a soda can, a stapler, and a spatula. In total, there were 12 targets consisting of nine go trials and three no go trials. The order and location of these targets was randomized across test days so participants did not learn sequences or target locations. In addition, different threatening/non-threatening items were used across the test days and were counterbalanced across participants. A go-pro camera mounted to the helmet of the participant served as a backup for scoring decision accuracy.

4.3 Human Factors and Marksmanship Metrics

Human factor's measures and questionnaires were completed throughout the scenario sequence. At intervals during the testing sessions, participants were asked to rate their level of perceived exertion using the Rating of Perceived Exertion (RPE) Scale [39]. The RPE Scale is a method for measuring perceived exertion and effort in physical work. The RPE is commonly determined in clinical diagnostics, therapy and rehabilitation, training of athletes and recreational sports. The RPE scale ranges from 6 to 20. A rating of 6 to 11 (very light) would essentially be the range for warm-up and cool-down. A rating of 12 to 13 is associated with approximately 60% of maximum heart rate and a rating of 16 to 20 is associated with approximately 90% of maximum heart rate. Before the start of a session and after volunteers finish each test session, they completed a rating of pain, soreness, or discomfort (RPSD) questionnaire [40] to indicate the level of discomfort experienced during the exercise. At the start of the assessment, each participant completed a Human Factors Demographic Questionnaire.

At intervals during the testing sessions (after rating perceived exertion on the RPE Scale), the participants were asked to rate the level of Interference/Degradation they experienced while performing a specific task due to the equipment configuration they were currently wearing. Additionally, tester observations and participants' comments were documented. At the conclusion of each test day, the participants completed a Human Factors questionnaire asking them to rate various aspects of the equipment configuration they wore that day. Specific attributes assessed included overall fit, overall comfort, ability to move/maneuver to accomplish mission activities, and overall acceptability for use in an operational environment.

Assessing the shooting performance generated five variables from the FN system data:

- The E-distance or variable error (VE) is a measure of the inconsistency (or consistency) in the outcome [41]. This measurement is the averaged Euclidian distance from the center of the shot series to each shot. If VE is small, the outcomes are consistent and located together closely. Another way to envision VE is "shot cluster size" or how tightly grouped the five shots in a shot series are.
- The B-distance or total variability (TV) is a measure of how accurately the shooter performed [41]. This measurement is the averaged Euclidian distance of a five-shot series to the bull's-eye. If the shot outcomes are close to the center of the target, TV is small. In other words, this variable is "shot accuracy."
- The total time between shots is calculated from time-stamps automatically recorded by the FN Expert Simulator software.
- Aiming Time is a variable automatically recorded by the FN Expert Simulator software. Aiming time represents the time a TP spends aiming at the target prior to firing (time starts when the system detects that the weapon is aimed at the target).
- Movement Time is calculated by subtracting the Aiming Time from the Total Time, and represents the time spent transitioning from one target to the next.

For the Go/No-Go task performance reaction time, accuracy, hit and false alarm rates were calculated from the go and no-go trials. In addition, d' and c were calculated (see formulas above). A false alarm is pressing to the infrequent no-go sound (M4 gunfire) and a hit is pressing correctly to the frequent go sound (AK47 gunfire). These values were then computed for each task block for each road march.

The same metrics described above were analyzed for marksmanship performance in the MOUT. In addition to these marksmanship measures, target response accuracy, hit and false alarm rates, and time to acquire the target are calculated. From the response accuracy data, response criterion and sensitivity can also be calculated. Sensitivity, d', is calculated using the following formula: $d' = z$ (proportion of hits) – z (proportion false alarms) and response criterion, c, is calculated as $c = -.5*$ (z (proportion of hits) +z (proportion of false alarms)). A false alarm is firing upon a non-threatening target and a hit is correctly firing upon a threatening target.

5 Conclusions

Our interdisciplinary and collaborative process that we have demonstrated is a comprehensive approach that addresses the Soldier as a system. The Soldiers' biomechanical, physiological, cognitive functions and physical performance are effected by the clothing and individual equipment that they wear. The paradigm of evaluating the Soldier from our approach allows the ability to analyze how these processes interact. This methodology gives a complete picture of what challenges Soldiers experience during complex scenarios. This research formulates and employs an evaluation methodology that establishes measurable standards for assessing Soldier performance in the next-generation of CIE.

Acknowledgments. The authors wish to thank the dedicated personnel from the NSRDEC, ATC, PDSSI, and Ft. Devens who made the conduct of our research possible. Special recognition is due to the study volunteers, who were enlisted men assigned to the 2nd Battalion, 504th Parachute Infantry Regiment, 82nd Airborne Division, Fort Bragg, NC.

References

1. Knapik, J., Harman, E., Reynolds, K.: Load carriage using packs: a review of physiological, biomechanical, and medical aspects. Appl. Ergon. **27**, 207–216 (1996)
2. Winsmann, F.R., Goldman, R.F.: Methods for evaluation of load-carriage systems. Percept. Mot. Skills **43**, 1211–1218 (1976)
3. Kennedy, S. J., Goldman, R., Slauta, J.: The carrying of loads within an infantry company. Technical report 73-51-CE. U.S. Army Natick Laboratories, Natick, MA (1973)
4. Department of the Army: Foot marches. Field Manual 21-18, Washington (1990)
5. Frykman, P.N., Harman, E.A., Knapik, J.J., Han, K.-H.: Backpack vs. front pack: differential effects of fatigue on loaded walking posture [Abstract]. Med. Sci. Sports Exerc. **26**, S140 (1994)
6. Hasselquist, L., Bensel, C.K., Corner, B., Gregorczyk, K.N.: An investigation of three extremity armor systems: Determination of physiological, biomechanical, and physical performance effects and quantification of body area coverage. Technical report NATICK/TR-12/014. U.S. Army Natick Soldier Research, Development and Engineering Center, Natick, MA (2012)
7. Harman, E., Han, K., Frykman, P., Pandorf, C.: The effects of backpack weight on the biomechanics of load carriage. Technical report T00-17. U.S. Army Research Institute of Environmental Medicine, Natick, MA (2000)
8. Martin, P.E., Nelson, R.C.: The effect of carried loads on the walking patterns of men and women. Ergonomics **29**, 1191–1202 (1986)
9. McLean, S., Perkins, N., Cain, S., McGinnis, R., Davidson, S., Vitali, R.: IMU array technology and algorithms for field-based warfighter performance assessment. Contractor's report W911QY-13-C-0011. U.S. Army Natick Soldier Research, Development and Engineering Center, Natick, MA (2015)
10. Cain, S.M., McGinnis, R.S., Davidson, S.P., Vitali, R.V., Perkins, N.C., McLean, S.G.: Using inertial measurement units to quantify gait performance. Paper presented at dynamic walking 2014, Zurich, Switzerland (2014)
11. Holewijn, M., Lotens, W.A.: The influence of backpack design on physical performance. Ergonomics **35**, 149–157 (1992)
12. Martin, P.E., Nelson, R.C., Shin, I.: Effects of backpack frame length, pack load, and participation time on the physical performance of men and women. Technical report NATICK/TR-82/021. U.S. Army Natick Research and Development Laboratories, Natick, MA(1983)
13. Knapik, J., Ang, P., Meiselman, H., Johnson, W., Kirk, J., Bensel, C., Hanlon, W.: Soldier performance and strenuous road marching: Influence of load mass and load distribution. Mil. Med. **162**(1), 62–67 (1997)

14. Harman, E., Frykman, P., Pandorf, C., Tharion, W., Mello, R., Obusek, J., Kirk, J.: Physiological, biomechanical, and maximal performance comparisons of female soldiers carrying loads using prototype U.S. Marine Corps Modular Lightweight Load-Carrying Equipment (MOLLE) with Interceptor body armor and U.S. Army All-Purpose Lightweight Individual Carrying Equipment (ALICE) with PASGT body armor. Technical report T99-9. U.S. Army Research Institute of Environmental Medicine, Natick, MA (1999)
15. Brainerd, S.T., Bruno, R.S.: Human factors evaluation of a prototype load-carrying system. Technical Memo 15-85. U.S. Army Human Engineering Laboratory, Aberdeen Proving Ground, MD (1985)
16. LaFiandra, M., Lynch, S., Frykman, P., Harman, E., Ramos, H., Mello, R.: A comparison of two commercial off the shelf backpacks to the Modular Lightweight Load Carrying Equipment (MOLLE) in biomechanics, metabolic cost and performance. Technical report T03-15. U.S. Army Research Institute of Environmental Medicine, Natick, MA (2003)
17. Pandorf, C.E., Nindl, B.C., Montain, S.J., Castellani, J.W., Frykman, P.N., Leone, C.D., Harman, E.A.: Reliability assessment of two militarily relevant occupational physical performance tests. Can. J. Appl. Physiol. **28**, 27–37 (2003)
18. Hasselquist, L., Bensel, C.K., Brown, M.L., O'Donovan, M.P., Coyne, M., Gregorczyk, K. N., Kirk, J.: Physiological, biomechanical, and maximal performance evaluation of medium rucksack prototypes. Technical report NATICK/TR-13/023. U.S. Army Natick Soldier Research, Development and Engineering Center Natick, MA (2013)
19. American College of Sports Medicine: ACSM'S resource manual for exercise testing and prescription, 6th edn. Lippincott, Williams & Wilkins, Baltimore (2010)
20. Byrne, N.M., Hills, A.P.: Relationships between HR and VO2 in the obese. Med. Sci. Sports Exerc. **34**, 1419–1427 (2002)
21. Dallack, L.C., Kravitz, L.: Relationships between % heart rate reserve and % VO2 reserve during elliptical cross trainer exercise. J. Sports Sci. Med. **5**, 662–667 (2006)
22. Lounana, J., Campion, F., Noakes, T.D., Medelli, J.: Relationship between % HR max, % HR reserve, % VO2 max and % VO2 reserve in elite cyclists. Med. Sci. Sports Exerc. **39**, 350–357 (2007)
23. Baca, A., Kornfeind, P.: Stability analysis of motion patterns in biathlon shooting. Hum. Mov. Sci. **31**(2), 295–302 (2012)
24. Hawkins, R.N., Sefton, J.M.: Effects of stance width on performance and postural stability in national-standard pistol shooters. J. Sports Sci. **29**(13), 1381–1387 (2011)
25. Tharion, W.J., Shukitt-Hale, B., Lieberman, H.R.: Caffeine effects on marksmanship during high-stress military training with 72 hours sleep deprivation. Aviat. Space Environ. Med. **74** (4), 309–314 (2003)
26. Warber, J.P., Tharion, W.J., Patton, J.E.: The effect of creatine monohydrate supplementation on obstacle course and multiple bench press performance. J. Strength Conditioning Res. **16**(4), 500–508 (2002)
27. Chang, Y.K., Labban, J.D., Gapin, J.I., Etnier, J.L.: The effects of acute exercise on cognitive performance: a meta-analysis. Brain Res. **9**, 1453–1487 (2012)
28. Lambourne, K., Tomporowski, P.: The effect of exercise-induced arousal on cognitive task performance: a meta-regression analysis. Brain Res. **1341**, 12–24 (2010)
29. Paas, F.G., Adam, J.J.: Human information processing during physical exercise. Ergonomics **34**, 1385–1397 (1991)
30. Tomporowski, P.D.: Effects of acute bouts of exercise on cognition. Acta Physiol. **112**, 297–324 (2003)
31. Tomporowski, P.D., Ellis, N.R.: Effects of exercise on cognitive processes: a review. Psychol. Bull. **99**, 338–346 (1986)

32. Crowell, H.P., Krausman, A.S., Harper, W.H., Faughn, J.A., Sharp, M.A., Mello, R.P., Smith, T., Patton, J.F.: Cognitive and physiological performance of Soldiers while they carry loads over various terrains. Technical report ARL-TR-1779. Army Research Laboratory, Aberdeen Proving Ground, MD (1999)
33. Krausman, A.S., Crowell, H.P., Wilson, R.M.: The effects of physical exertion on cognitive performance. Technical report ARL-TR-2844. Army Research Laboratory, Aberdeen Proving Ground, MD (2002)
34. Mastroianni, G.R., Chuba, D.M., Zupan, M.O.: Self-pacing and cognitive performance while walking. Appl. Ergon. **34**, 131–139 (2003)
35. Dietrich, A.: Functional neuroanatomy of altered states of consciousness: the transient hypofrontality hypothesis. Conscious. Cogn. **12**, 231–256 (2003)
36. Dietrich, A., Audiffren, M.: The reticular-activating hypofrontality (RAH) model of acute exercise. Neurosci. Biobehav. Rev. **35**, 1305–1325 (2011)
37. Morris, T., Sproule, J., Turner, A., Hale, B.J.: Acute, intermediate intensity exercise, and speed and accuracy in working memory tasks: a meta-analytical comparison of effects. Physiol. Behav. **102**(3–4), 421–428 (2011). doi:10.1016/j.physbeh.2010.12.007
38. Eddy, M., Hasselquist, L., Howe, J., Rourke, J., Coyne, M., O'Donovan, M., Batty, J., Brunye, T., Mahoney, C.: The effects of load carriage and physical fatigue on cognitive performance. PLoS ONE **10**(7), e0130817 (2015)
39. Borg, G.A.V.: Perceived exertion as an indicator of somatic stress. Scand. J. Rehabil. Med. **2**, 92–99 (1970)
40. Corlett, E., Bishop, R.: A technique for assessing postural discomfort. Ergonomics **19**, 175–182 (1976)
41. Schmidt, R.A., Lee, T.D.: Motor control and learning: a behavioral approach. Human Kinetics, Champaign (1999)

Dynamic Model of Athletes' Emotions Based on Wearable Devices

Damien Dupré[1(✉)], Ben Bland[2], Andrew Bolster[2], Gawain Morrison[2], and Gary McKeown[1]

[1] School of Psychology, Queen's University, Belfast BT7 1NN, UK
{Damien.Dupre,G.McKeown}@qub.ac.uk
[2] SensumCo. Ltd., Belfast BT9 6RY, UK
{Ben,Bolster,Gawain}@sensum.co

Abstract. With the development of wearable sensors, it is now possible to assess the dynamic progression of physiological rhythms such as heart rate, breathing rate or galvanic skin response in ways and places that were previously impractical. This paper presents a new application that synchronizes the emotional patterns from these time-series in order to model athletes' emotion during physical activity. This data analysis computes a best-fitting model for analyzing the patterns given by these measurements "in the wild". The recording setup used to measure and synchronize multiple biometric physiological sensors can be called a BAN (Body Area Network) of personal measurements. By monitoring physical activity, it is now possible to calculate optimal patterns for managing athletes' emotion. The data provided here are not restricted by a lab environment but close to the "ground truth" of ecologically valid physiological changes. The data allow the provision of accurate feedback to athletes about their emotion (e.g. in cases such as an unexpected increase or an expected decrease of physiological activity).

Keywords: Physiology · Wearable sensors · Multimodal measure · Outdoor activities · Time-series

1 Role of Emotion During Athletes' Performance

Performance monitoring is an increasingly important factor for competing in elite sports. It is essential not only from an individual perspective – supporting the athlete's autonomy in devising individualized training programs – but also at a collective level in helping the athlete's support network to understand the athlete's performance. Psychological performance indicators have typically been underestimated or overlooked, often due to their lack of reliability, even though the importance of mental conditioning is commonly acknowledged. However, psychological assessment is increasingly taken more seriously, across both athletes and their teams.

Over longer periods, athletes' performance in sports and activities is not stable. Performance increases and decreases according to both physical and psychological states. In light of this, a large range of sports have been investigated in a series of studies (e.g. high endurance trail [1], motor car racing [2], cycling [3] and parachute

T. Ahram (ed.), *Advances in Human Factors in Sports, Injury Prevention and Outdoor Recreation*, Advances in Intelligent Systems and Computing 603,
DOI 10.1007/978-3-319-60822-8_4

jumping [4]; the suggested references provide useful resources but are not connected to the current series of studies). The current paper reports on the method, analysis, and results exploring athlete's emotional responses during a mountain biking study that was part of the larger series of studies.

Among the numerous variables influencing an athlete's psychological state, emotions seem to play a decisive role [5, 6]. Aristotle described the concept of Eudaemonia – a state of happiness and wellbeing that results from reaching the highest possible manifestation of one's nature. To reach this state, athletes need to grow through personal experiences. This growth often involves periods of intense emotion, conquering peaks of anxiety, fear and terror to reach triumph, exhilaration and relief. The performance details this journey, allowing athletes to live through a hero's passage from the negative to the positive, from fear to joy. The word hero is not used lightly it is often this narrative and emotional commitment behind the observable sport that compels interest in any given athlete's progression and the sport itself.

An athlete's emotions can be thought of as a multi-componential process [7]. This process simultaneously involves cognitive appraisal [8], expressive patterns [9], action tendencies [10], subjective feeling [11] and physiological changes [12, 13]. This last component of emotions appears to be particularly relevant in the case of athlete performance. Physiological changes seem to prepare the body to react to a situation, classically preparation for "fight or flight". Understanding, influencing and regulating their emotions allows athletes to aim for better results and performance. Evaluating physiological responses is not only used to assess athletic performance but also to inform military and health issues [14, 15]. These systems described by what we can call the Body Area Network (BAN).

2 Multimodal Body Area Network for Athletes

Even if measuring emotions is possible in lab conditions, it remains a challenge in elite or extreme sports environment. Data signals can be disrupted by all the added "noise" of exertion and movement, such as when the athlete would already have an elevated breathing and heart rate. Then there are additional problems from the athlete's environment, with extreme conditions (e.g. temperature, vibration, speed, G-force). In order to evaluate athletes' emotions in the wild, an experiment was performed with a range of biometric sensors – and media sensors for context – all synchronized through a purpose-built "SYNC" app (designed by Sensum). The app was connected wirelessly to the biometric sensors, cameras and microphones, triggering them simultaneously and capturing their data.

The SYNC app measures emotions by integrating data from third-party wearables such as smartwatches and fitness bands, medical devices like skin conductance and heart rate sensors, as well as cameras, microphones and other devices and methods. Sensum tests all this equipment and sets baselines for measuring emotions with biometric sensors. In this study we were looking for a floor value, from which emotional data can rise above the noise of the situation.

3 Method

In order to evaluate the physiological changes of mountain bikers during their runs on downhill trails, a Body Area Network was created for each of four riders, capturing individual measurements as they performed six runs on each of two different trails.

3.1 Participants

Four riders were recruited for this experiment (Table 1). They were recruited according to their familiarity with the selected trails. They had already navigated these trails many times and were very familiar with the different obstacles they presented. Before their participation, the riders completed informed consent forms which noted that they could leave the experiment at any time.

Table 1. General demographic description of the four riders

ID	Gender	Age	Mountain bike general level
Rider 1	Male	14	Semi-professional
Rider 2	Female	16	Advanced
Rider 3	Male	19	Professional
Rider 4	Male	39	Professional

3.2 Measurements

For this experiment, the riders wore an Equivital EQ02 sensor belt, a GoPro Hero 4 camera and a Sennheiser MKE 2 lavalier microphone with inbuilt Apogee digital signal converter. They were also wearing an iPhone on their right arm (for recording audio from the lavalier mics via an Apogee app) and an OnePlus X Android smartphone on their left arm (for Sensum app to control sensors) in order to record the signals given by the wearable devices (Fig. 1). The Android smartphone also provided GPS coordinates and three-axis accelerometer/magnetometer data.

Over the available wearable devices, the Equivital EQ02 sensor belt was chosen based on the accuracy and robustness of its measurement. The description of the Equivital measurement is provided in the Table 2.

The measures recorded can be divided into two categories: the physiological measures: HR = Heart Rate (BPM), BR = Breathing Rate (RPM), GSR = Galvanic Skin Response (μS), ST = Skin Temperature (°C); and the context measurements (Speed, Acceleration and perceived difficulty). The perceived difficulty of the trails was evaluated afterward by the most experienced rider using a dynamic scale from 0 (no difficulty) to 100 (hardest difficulty) while he was watching his own point-of-view GoPro video recording. These context measures will be used as predictors of the physiological changes. There are less measurements for Galvanic Skin Response as our ethical protocol provided that the Riders would only wear a sensor if they were fully comfortable and this decision was the Rider's alone. Rider 1 chose not to wear the GSR equipment and Rider 2 and 3 only did so on a subset of the runs.

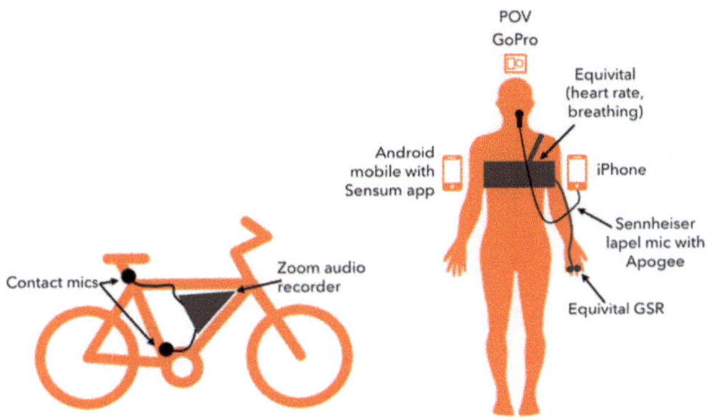

Fig. 1. Description of the BAN setup. POV = point-of-view, GSR = galvanic skin response.

Table 2. Description of the Equivital EQ02 sensors.

Signal	System required	Sample rate
Electrocardiogram (ECG - 2 Channels)	Core	256 Hz
Heart Rate (HR)	Core	25.6 Hz
Breathing Rate (BR)	Core	25.6 Hz
Accelerometer (ACC - 3 Axis)	Core	25.6 Hz
Skin Temperature (ST)	Core	1/15 s
Galvanic Skin Response (GSR)	Core + GSR Ancillary	2 Hz

3.3 Procedure

The experiment took place in the Rostrevor Forest/Kilbroney Park, Shore Road, BT34 3AA, Northern Ireland (www.mountainbikeni.com/rostrevor/).

Two different official mountain-bike trails were selected for this experiment (Fig. 2). The first one, called "Mega Mission", is 1.8 km long. It is made of big tabletop and hip jumps, and rhythm sections that allow riders to pick optional lines and jumps depending on skill level, making it the faster, smoother trail with more aerial time. The second trail, called "On the Pulse," is 1.3 km long. It includes several "rock gardens" (rocky sections) that lead into some back-to-back "berms" (banked corners). The average speed of this trail is not as fast as Mega Mission but it is considered more technical, mainly due to the rock gardens.

The experiment was performed over the course of three days. During this time, the riders were asked to perform each trail 6 times: 12 runs in total (2 in the morning and 2 in the afternoon). Each Rider was separated by three minutes' interval to ensure enough time to avoid them disrupting each other's run.

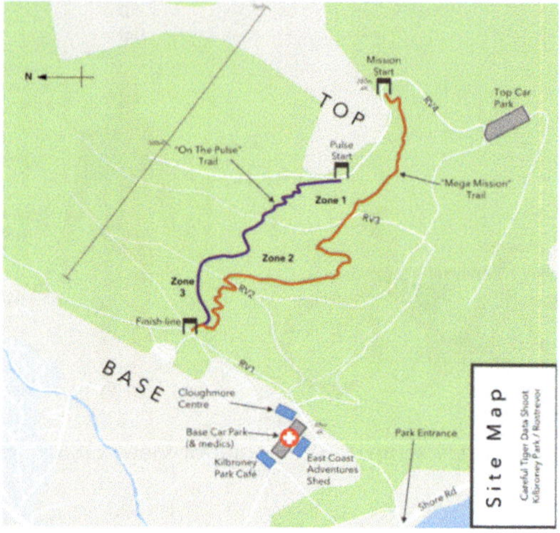

Fig. 2. Geographic layout of the selected Trails, "Mega Mission" in orange and "On the Pulse" in purple.

4 Results

To evaluate the influence of context on the rider's physiological responses, we first present a descriptive statistical analysis, before assessing an emotion score for arousal and valence derived from the rider's biometrics, and correlating these scores against contextual measurements of difficulty, speed, and acceleration.

Descriptive Results. As expected, physiological measurements vary with individual and contextual variables. From descriptive statistics, we observe an important variation according to the Riders and the Trail performed (Fig. 3).

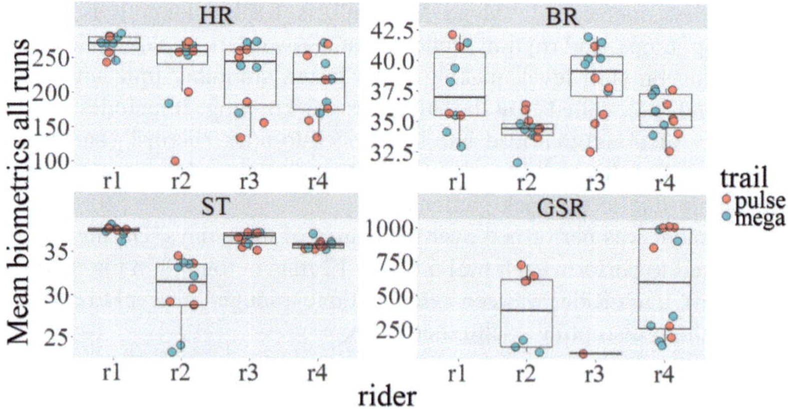

Fig. 3. Mean of the biometrics measured by the Equivital EQ02 sensors represented by the dots according to rider and trail type. The box plot is an indicator of their mean distributions (median and quartiles).

For Heart Rate, Breathing Rate and Galvanic Skin Response, we can observe differences both between the different riders and within the individual riders' data for the different courses. For Breathing Rate and Skin Temperature, Rider 2 shows a different range of values to the other riders. Galvanic Skin Response was the only biometric measurement that showed a marked difference between each trail, however the reduced amount of data makes generalization more difficult. Note that ECG descriptive statistics are not presented here because raw ECG data are not meaningful without treatment.

Emotion Analysis. We adopted a classic two dimensional (arousal and valence) model of emotions [16], arousal and valence were interpreted from the biometric time-series of all the riders and plotted according to the GPS coordinates of the trails (Figs. 4 and 5). The arousal and valence scores are varied according to the degree of saturation of the biometric components (0 indicates a neutral position).

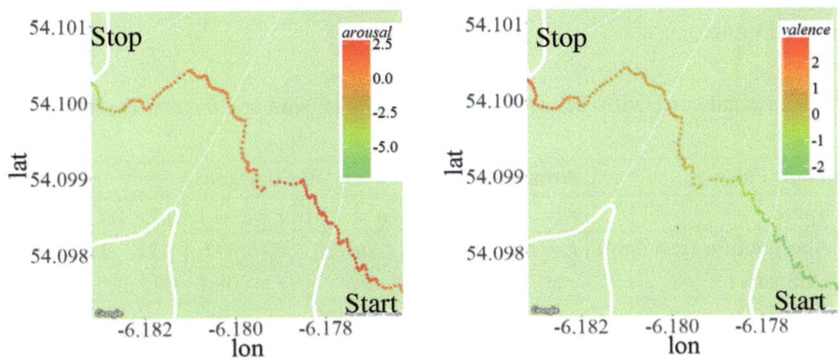

Fig. 4. Evolution of emotional arousal (left) and valence (right) according to the GPS coordinates of the Pulse trail.

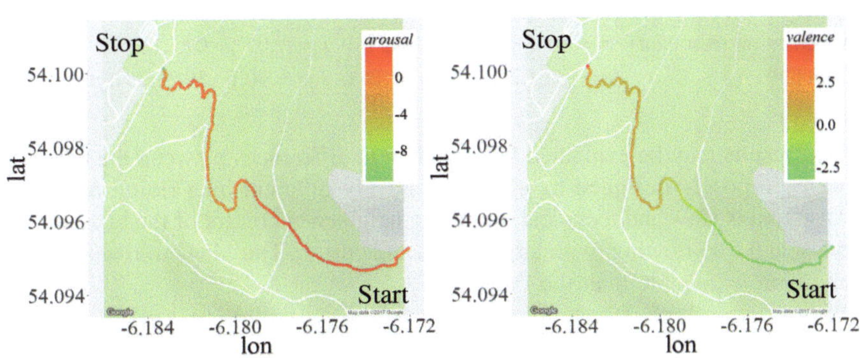

Fig. 5. Evolution of emotional arousal (left) and valence (right) according the GPS coordinates of the Mega trail.

The results from the emotion analysis shows that riders typically experienced the highest arousal state at the beginning of each trail, however, arousal levels remained relatively high throughout the trail, presumably due to the presence of challenging obstacles throughout each trail. The level of arousal dropped rapidly at the end of the trail. In contrast, valence measurements showed that they started of their runs in a negatively valenced state and they mainly experienced positive emotions at the end of their runs. The change from negative to positive valence proceeded through a relatively smooth gradient as the run progressed.

Contextual Influences. The correlation between emotional dimensions and contextual measures shows significant differences between the two trails (Tables 3 and 4). On the Pulse trail, the arousal dimension is only correlated with speed whereas the valence dimension does not appear to be correlated with any contextual measure. However, on the Mega trail arousal and valence are both correlated with the rider's speed, and with the perceived difficulty of the trail, whereas riders' acceleration is not correlated with emotional dimensions.

Table 3. Temporal correlation between the emotion dimension and the contextual measures for the Pulse trail.

Trail: Pulse	Arousal	Valence
Speed	$r = .42$ [.30, .54], $p < .001$	$r = .04$ [−.11, .18], *ns.*
Acceleration (resultant)	$r = .07$ [−.08, .21], *ns.*	$r = −0.6$ [−.21, .08], *ns.*
Difficulty	$r = .14$ [−.01, .28], *ns.*	$r = .08$ [−.07, .22], *ns.*

Table 4. Temporal correlation between the emotion dimension and the contextual measures for the Mega trail.

Trail: Mega	Arousal	Valence
Speed	$r = .59$ [.50, .67], $p < .001$	$r = −.28$ [.40, .16], $p < .001$
Acceleration (resultant)	$r = .01$ [−.11, .14], *ns.*	$r = −.05$ [−.07, .18], *ns.*
Difficulty	$r = .35$ [.23, .45], $p < .001$	$r = −.40$ [−.50, −.29], $p < .001$

These results may be explained by the inherent differences between the two trails – they were purposely designed to give qualitatively different bike riding experiences. The Pulse trail is slow and technical, requiring heightened control of the bike, the Mega trail is fast and free-flowing, with large jumps and curves. The Mega trail was described as being more enjoyable than the Pulse trail.

5 Discussion

Evaluating athletes' emotions is a way to understand the factors influencing their performance. It can maximize the capacity of each athlete to analyze his or her results. Emotion measurement can tell an athlete what their body is experiencing, potentially providing an alternative indicator affecting performance, compared to contextual measurements such as speed or acceleration. Appropriate measurement of emotions may even demystify and enable regulation of a previously difficult to address set of performance related variables. Different stages of the emotional experience are measurable, from the anxious build-up, through the terrifying climax, to the relieving aftermath. Understanding these may allow greater control and improve performance.

In performance-critical environments it is valuable to measure the situational data, and emotion data in particular, and make them applicable to the athlete's performance and training programs. Sensum's SYNC app records and analyzes emotions and it also supports the interpretation and communication of that data, transforming it into visual and concise information to help athletes measure and understand their emotions.

Emotion as athletic biofeedback opens a new perspective in understanding athletes' motives and can be useful in supporting performance not just in competition but more generally to inform the athlete's emotional development and social support [17]. An interesting research avenue will be investigating how athletes' emotion biofeedback can help them to modify their appraisal of their situation [18].

From contextual measurements, it is possible to categorize an athlete's style (e.g. using runner's elevation velocity and steps-per-minute in Ultramarathons [1]) but measuring psychological variables, such as emotion, to map a personal representations and to identify individual affective style, offers an important new perspective in understanding athletic performance and the motivation drivers to perform [19], and equally it may provide new explanatory variables when underperformance is observed.

However, emotion measurements for performance assessment bring the same problems as existing biometrics used in sports assessment [20]. The main concern is the validity and interpretation of data, which if incorrectly applied can lead to false analysis. Therefore to minimize errors, the emotional model under scrutiny in this paper will have to be tested and validated in different situations to test its generalizability and robustness. A significant improvement would be the removal of incoherent data due to sensor error (such as heart rates greater than 220 BPM or skin temperatures lower than 28 °C). Another important ethical concern is the increase of surveillance and threats to privacy that are inherent in the pervasive use of these new technologies. These data are obviously personal and must be protected from abuse and only used to increase their performance and wellbeing.

Nevertheless, wearable technologies show great promise in providing rich physiological measurements and thus monitoring athletes' performance. They are particularly interesting in evaluating the variability of the physiological rhythms that occur during athletes' training and performance experiences.

References

1. Hauw, D., Rochat, N., Gesbert, V., Astolfi, T., Philippe, R.A., Mariani, B.: Putting together first-and third-person approaches for sport activity analysis: the case of ultra-trail runners' performance analysis. In: Advances in Human Factors in Sports and Outdoor Recreation, pp. 49–58. Springer (2017)

2. Schwaberger, G.: Heart rate, metabolic and hormonal responses to maximal psycho-emotional and physical stress in motor car racing drivers. Int. Arch. Occup. Environ. Health **59**(6), 579–604 (1987)
3. Lucia, A., Hoyos, J., Carvajal, A., Chicharro, J.: Heart rate response to professional road cycling: the Tour de France. Int. J. Sports Med. **20**(03), 167–172 (1999)
4. Cavalade, M., Papadopoulou, V., Theunissen, S., Balestra, C.: Heart rate variability and critical flicker fusion frequency changes during and after parachute jumping in experienced skydivers. Eur. J. Appl. Physiol. **115**(7), 1533–1545 (2015)
5. Hetland, A.: Feeling the extreme: an exploratory study of experienced emotions during extreme sport. University of Tromsø, Norway, Master thesis (2009)
6. Hetland, A., Vittersø, J.: The feelings of extreme risk: exploring emotional quality and variability in skydiving and BASE jumping. J. Sport Behav. **35**(2), 154–180 (2012)
7. Scherer, K.R.: The dynamic architecture of emotion: evidence for the component process model. Cogn. Emot. **23**(7), 1307–1351 (2009)
8. Scherer, K.R.: Appraisal theory. In: Dalgleish, T., Power, M.J. (eds.) Handbook of Cognition and Emotion, pp. 637–663. Wiley, New York (1999)
9. Russell, J.A., Fernandez-Dols, J.M.: The Psychology of Facial Expression. Cambridge University Press, Cambridge (1997)
10. Frijda, N.H.: Emotion, cognitive structure, and action tendency. Cogn. Emot. **1**(2), 115–143 (1987)
11. Damasio, A., Carvalho, G.B.: The nature of feelings: evolutionary and neurobiological origins. Nat. Rev. Neurosci. **14**(2), 143–152 (2013)
12. Kreibig, S.D., Samson, A.C., Gross, J.J.: The psychophysiology of mixed emotional states. Psychophysiology **50**(8), 799–811 (2013)
13. Kreibig, S.D.: Autonomic nervous system activity in emotion: a review. Biol. Psychol. **84** (3), 394–421 (2010)
14. Holliman, J., Zhivich, M., Khazan, R., Swiston, A., Telfer, B.: Building low-power trustworthy systems: cyber-security considerations for real-time physiological status monitoring. Presented at the Military Communications Conference, pp. 1083–1089 (2016)
15. Rushanan, M., Rubin, A.D., Kune, D.F., Swanson, C.M.: SoK: security and privacy in implantable medical devices and body area networks. Presented at the Symposium on Security and Privacy, pp. 524–539 (2014)
16. Mehrabian, A., Russell, J.A.: An Approach to Environmental Psychology. MIT Press, Cambridge (1974)
17. Fuller, M.S.: 'Hey, Nice Run!' Exploring Motives for Smartphone Exercise App Use, University of Connecticut (2016)
18. Bartholomew, K.J., Arnold, R., Hampson, R.J., Fletcher, D.: Organizational stressors and basic psychological needs: the mediating role of athletes' appraisal mechanisms. Scand. J. Med. Sci. Sports **16**, 1–13 (2017)
19. Apicella, C., Demiral, E.E., Mollerstrom, J.: No Gender Difference in Willingness to Compete When Competing Against Self. DIW Berlin, Berlin (2017)
20. Karkazis, K., Fishman, J.R.: Tracking US professional athletes: the ethics of biometric technologies. Am. J. Bioeth. **17**(1), 45–60 (2017)

Injury Prevention and Analysis of Individual and Team Sports

Design of a Secure Biofeedback Digital System (BFS) Using a 33-Step Training Table for Cardio Equipment

Xiaokun Yang[1(✉)] and Nansong Wu[2]

[1] College of Science and Engineering, University of Houston Clear Lake, 2700 Bay Area Blvd, Hoston, TX 77058, USA
yangxia@uhcl.edu

[2] Department of Electrical Engineering, Arkansas Tech University, 1811 N Boulder Ave, Russellville, AR 72801, USA
nwu@atu.edu

Abstract. In order to efficiently instruct aerobic training on cardio equipment, this paper proposes a secure biofeedback digital system (BFS) with a 33-step training table. As a case study, the system is simulated using Virtex5-110t filed-programming gate array (FPGA) to run-time monitor body information and feedback guide the exercise intensity. Experimental results show that the BFS system can be effectively implemented with 11352 slices cost and 618 mW dynamic power consumption. Additionally, the throughput can reach 2.30 Gbps for cipher tests.

Keywords: Aerobic training · Biofeedback digital system · Cardio equipment · Filed-Programming Gate Array (FPGA) · Power consumption

1 Introduction

From proper aerobic exercise human bodies will derive a very efficient cardiovascular system (the heart, lungs, and circulation), very high levels of bone and joint support and body-fat burning, as well as other benefits. These benefits can be derived from many types of exercises, including walking, stationary biking, aerobic running, and others, when simple guidelines are followed [1, 2]. One of the most important aspects of the exercise program is reading body information and analyzing training data in real time to improve the exercise performance and minimize risk of injuries. A much easier approach, one that is relatively simple and accurate since it parallels laboratory data, is to use a smart and Internet-connected cardio equipment to feedback regulate training intensity and analyze exercise performance.

Today, much of the research, such as the use of inertial sensors to biofeedback control the treadmills [3, 4] and estimate the energy expenditure [5, 6], has been proposed. While all these approaches mainly focus on the prediction accuracy and power consumption of systems running on hardware equipment. They apply ready-to-use microcontrollers to build up the training systems. The controllability of data privacy and power consumption offered by such embedded chips is rather limited.

T. Ahram (ed.), *Advances in Human Factors in Sports, Injury Prevention and Outdoor Recreation*, Advances in Intelligent Systems and Computing 603,
DOI 10.1007/978-3-319-60822-8_5

More important, the commonly used hardware structures, such as the AMBA Advanced High-Performance Bus (AHB) [7], Advanced eXensible Interface (AXI) [8], Wishbone [9], and OCP [10], are very complicated in order to provide a broad range of applications. Thus, these circuits are much more costly in terms of both area and energy consumption. To tackle this issue, a low-cost and low-power bus protocol is proposed in [11], which is suitable to the small-scale and resource-limited hardware designs. Moreover, an advanced structure is further presented in [12], in order to improve the data feeding efficiency for Advanced Encryption Standard (AES) [13], since security and privacy is becoming the top tier concern for such Internet-connected things and leading up to a significant share of system resources.

Based on the aforementioned problems, this paper proposes an application-specific circuit using [11, 12] hardware structure, in order to run-time regulate workout intensity to maximize exercise economy, and analyze the training performance based on a 33-step aerobic training table. The main contributions are as follows.

(1) We presented a high-performance AES-encrypted system to balance hardware performance with limited resource on small-size circuits.
(2) We proposed a novel 33-step aerobic training table as an extension of the maximum aerobic function (MAF) theory [1], aiming to guide exercise intensity to a desired level through regulating heart rate.
(3) We implemented a hardware module to provide exercise data, involving calorie consumption, workout time, distance, user ID, and MAF training level, to show the training performance.
(4) We designed a platform using both MSBUS and AXI structures, involving register-transfer level (RTL) design, verification, synthesis, place and route, and power analysis, to evaluate the system performance.

The remainder of this paper is organized as follows: Sect. 2 introduces the 33-step aerobic training, and Sect. 3 presents the biofeedback circuit design, including the calculation module and AES engine. Then, the experimental results are summarized in Sect. 4. Finally, Sect. 5 concludes this paper.

2 33-Step Aerobic Training

In this section, we present the 33-step aerobic training table, and then briefly explain the training procedure.

2.1 33-Step Aerobic Training

Since workout intensity directly affects heart rate, a heart rate monitoring becomes an important application in quantifying workout. To find the maximum aerobic heart rate (MA-HR), there are two important steps introduced in [1]:

(1) First, subtract the age from 180, which is defined as 180-formula.
(2) Next, find the best category for the present state of fitness and health, and make the appropriated adjustments. Then, try to keep the heart rate as MA-HR during exercises and measure the average pace.

In our study, we extended the MAF theory and presented a novel 33-step aerobic training table for testing a runner's maximum aerobic function. We performed the test on a treadmill, where 5, 1-kilometer splits at the MA-HR are recorded in Table 1. The level of exercise intensity is one of the main factors building a strong and fast aerobic system. By setting up this system, fat burning increases, physical injuries can be prevented, and running economy can improve significantly.

Table 1. 33-step aerobic training table.

33-level	5k-MAF pace	5k-race pace	5k time
33	3:15	2:40	13:20
32–31	3:25–3:35	2:50–3:00	14:10–15:00
30–28	3:45–4:15	3:10–3:30	15:50–17:30
27–25	4:30–5:00	3:40–4:00	18:20–20:00
24–21	5:15–6:15	4:10–4:40	20:50–23:20
20–17	6:35–7:35	4:50–5:20	24:10–26:40
16–13	7:55–8:55	5:30–6:00	27:30–30:00
12–9	9:25–11:55	6:15–7:00	31:15–35:00
8–5	12:55–15:33	7:30–9:00	37:30–45:00
4–1	20:15–17:00	11:00–9:30	55:50–47:30
0	21:20	11:30	57:30

More specifically, in the first column we divide the MAF level into 33 steps according to the 5k-MAF pace, from the normal walking speed to the world record of the 5k running race. Notice that step 0 to step 32 are defined for female, and step 1 to step 33 are defined for male. The MAF pace shown in the second column is directly related to the competitive performance, such as the race pace and race time shown in the third and fourth columns.

It is important to objectively assess the progress of the aerobic system, or more specifically that of 5k-MAF speed. Using basic biofeedback, one's MAF level should gradually improve as indicated by running faster at the same MA-HR. This evaluates a runner's aerobic function and becomes one of the most important assessment tools.

2.2 33-Step Training Procedure

In what follows, the 33-step training procedure is shown in Fig. 1. The first step is loading users' exercise information such as age in years (A), weight in kilograms (W), exercise duration time in minutes (T), running distance in kilometers (D), and user ID, and also the previous training data involving calorie consumption (Cal) and the MAF training level (Lel). Then, the MA-HR can be computed using the 180-formula and the 5k-MAF pace can be matched using Table 1.

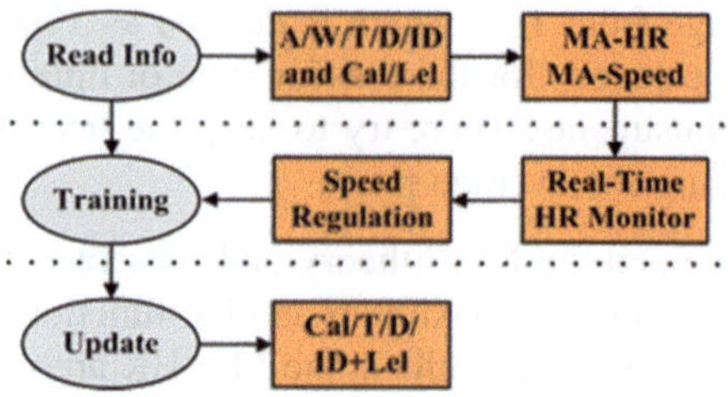

Fig. 1. MAF-based training procedure.

In the second step, the heart rate monitoring is employed as a coach, helping users to control their activities more precisely. The MAF training normally comprises of three stages: warm-up from the resting heart rate (RHR) to MA-HR gradually, aerobic training under MA-HR, and cool-down from the MA-HR to RHR. This process can be automatically regulated on cardio equipment.

Finally, the new training data, including calorie consumption, exercise time, running distance, and training level, are computed, encrypted, and updated to users' account. They can be applied to test training performance and optimize the future workout plan.

3 Hardware System Implementation

This section presents the system design using the 33-step training theory. In our work, both MSBUS and AXI structures are applied to evaluate the system performance.

3.1 System Environment

As shown in Fig. 2, the design-under-test (DUT) includes a calorie calculation module and a security module. All the other models such as the micro-processor and heart rate monitor are verification intellectual properties (IPs).

As the only master of the control bus (MBUS), the micro-processor configure all the application-specific modules and peripherals. Basically, it can monitor the bio-status through timely processing the interrupts, and communicate with embedded systems on cardio equipment through SPI and UART.

As the only slave of the data bus (SBUS), the calculation module can be accessed by all the SBUS masters. 4-byte data, including 1-byte calorie burning, 1-byte exercise time, 1-byte exercise distance, as well as 3-bit user ID and 5-bit training level, are encrypted/decrypted as one column of an AES state.

3.2 Calculation Module Design

A circuit with heart rate monitoring feature is an important tool that not only guides the training, but also is part of an important assessment process, and can even be used in

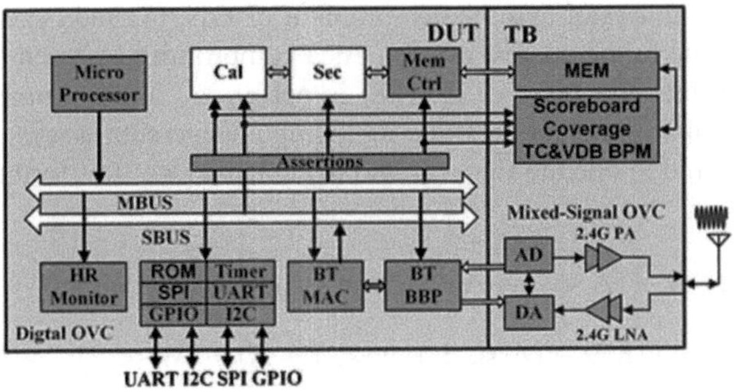

Fig. 2. System environment.

some applications, such as estimating the number of calories burned during exercise. The calorie burning formulas for male and female athletes [6], respectively, are

$$\text{MKcal} = [(-55.0969 + (0.6309 \times \text{HR}) + (0.1988 \times \text{W}) + (0.2017 \times \text{A})/4.184] \times \text{T} \quad (1)$$

$$\text{FKcal} = [(-20.4022 + (0.4472 \times \text{HR}) - (0.1263 \times \text{W}) + (0.074 \times \text{A})/4.184] \times \text{T} \quad (2)$$

Ignoring the run-time changed heart rate, the equations can be simplified as

$$\text{MKcal} = [(58.4664 - (0.4292 \times \text{A}) + (0.1988 \times \text{W})/4.184] \times \text{T} \quad (3)$$

$$\text{FKcal} = [(60.0958 - (0.3732 \times \text{A}) - (0.1263 \times \text{W})/4.184] \times \text{T} \quad (4)$$

Although the calorie burning computation using Eqs. (3) and (4) is not as accurate as the result of Eqs. (1) and (2), the system power consumption can be largely reduced without collecting the run-time modified heart rate.

In addition, scientific algorithms are typically developed in high-level languages such as Matlab or C. Single- or double-precision floating-point data are used to support wide dynamic range while maintaining data precision. However, full floating-point arithmetic operators dramatically increase logic utilization and energy consumption, as well as low speed compared with fixed-point designs [14]. Instead of just throwing more or larger logic circuits at the problem, therefore, we convert the floating-point Eqs. (3) and (4) to fixed-point formulas as follows

$$\text{MKcal} = [(1789 - (13 \times \text{A}) + (6 \times \text{W})]/128 \times \text{T} \quad (5)$$

$$\text{FKcal} = [(1839 - (11 \times \text{A}) - (4 \times \text{W})]/128 \times \text{T} \quad (6)$$

where both of the numerator and the denominator of Eqs. (3) and (4) are multiplied by 30.6 and rounded up to the nearest integer. As a compromise between accuracy and bit width utilization, the 7th power of 2 (128) is performed as the numerator. After the conversion, several arithmetic operators, including adders, subtractors, multipliers, and dividers, are required to operate Eqs. (5) and (6) in hardware. The further simplification can be rewritten as

$$\mathrm{MKcal} = [(1789 - (\mathrm{A} \ll 3 + \mathrm{A} \ll 2 + \mathrm{A}) + (\mathrm{W} \ll 2 + \mathrm{W} \ll 1)] \gg 7 \times \mathrm{T} \quad (7)$$

$$\mathrm{FKcal} = [(1839 - (\mathrm{A} \ll 3 + \mathrm{A} \ll 1 + \mathrm{A}) + (\mathrm{W} \ll 2)] \gg 7 \times \mathrm{T} \quad (8)$$

Here, the shift operators are used to replace multipliers and dividers. The shift operators "$\ll$" and "$\gg$", respectively, perform left and right shift of their left operand by the number of bit positions given by the right operand.

Finally, 5 adders and 1 multiplexer are required to compute the calorie burning. In order to improve the maximum operational frequency, we split the combinational part into two cycles to shorten the critical path. Meantime, the training level can be read out by matching up the average pace in Table 1.

3.3 AES Core Implementation

Security and privacy is becoming a de-facto requirement of small-size integrated circuits. Hence, we design and integrate an AES engine in our system with a 10-round algorithm using $\emptyset = \{10\}_2$ and $\lambda = \{1100\}_2$ [15].

Basically, each AES round except for the final round consists of four different byte-oriented transformations: non-linear byte substitution using a S-box (SB/ISB), shifting rows of the state array (SR/ISR), mixing the data within each column of the state array (MC/IMC), and adding a round key to the state (AK), while the final round does not have the MC/IMC transformation. Figure 3 shows a 32-bit AES system structure including encryption (ENC) and decryption (DEC) cores. For the non-cipher tests of this system, both ENC and DEC engines can be bypassed. Otherwise, the write data should be decrypted before being stored into the memory, and the read data should be encrypted before being transferred on the data bus.

Considering the system speed, we split both the AES engine into sub-stage1 and sub-stage2 (SS1 and SS2). The SB/ISB transformation is decomposed into a modular inversion over GF(2 ^ 4) and four linear functions (A, IA, δ, and Iδ). In order to shorten the critical path of SB/ISB, IA is combined with δ (IA × δ) in sub-stage1, and Iδ is merged with A (Iδ × A) in sub-stage2. In addition, the SR/ISR, MC/IMC, and AK transformations are integrated in sub-stage2 to obtain approximately equal delay to sub-stage1.

Generally, all the operators such as δ and Iδ shown in Fig. 3 can be imagined as black boxes with logic input "a" and output "b". The bit width is 8-bit, 4-bit, and 2-bit,

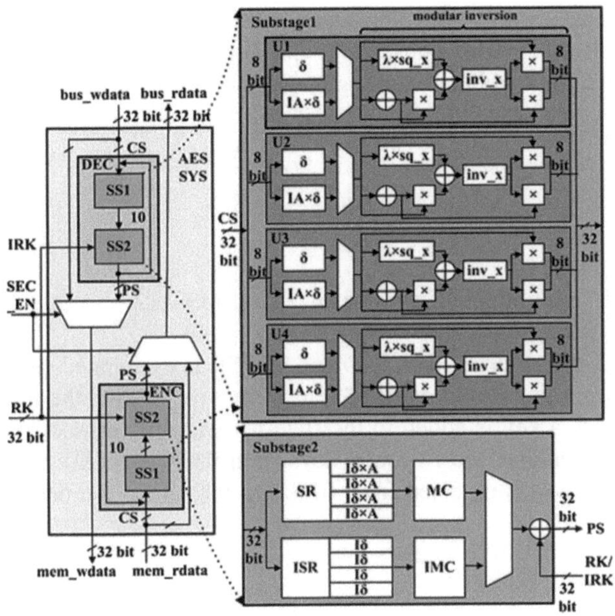

Fig. 3. AES core structure.

respectively, in GF(2 ^ 8), GF(2 ^ 4), and GF(2 ^ 2) fields. For instance, the logic of δ and Iδ can be written as

$$b = \{a_7 \wedge a_5, a_7 \wedge a_6 \wedge a_4 \wedge a_3 \wedge a_2 \wedge a_1, a_7 \wedge a_5 \wedge a_3 \wedge a_2, a_7 \wedge a_5 \wedge a_3 \wedge a_2 \wedge a_1 \\ a_7 \wedge a_6 \wedge a_2 \wedge a_1, a_7 \wedge a_4 \wedge a_3 \wedge a_2 \wedge a_1, a_6 \wedge a_4 \wedge a_1, a_6 \wedge a_1 \wedge a_0\} \tag{9}$$

$$b = \{a_7 \wedge a_6 \wedge a_5 \wedge a_1, a_6 \wedge a_2, a_6 \wedge a_5 \wedge a_1, a_6 \wedge a_5 \wedge a_4 \wedge a_2 \wedge a_1 \\ a_5 \wedge a_4 \wedge a_3 \wedge a_2 \wedge a_1, a_7 \wedge a_4 \wedge a_3 \wedge a_2 \wedge a_1, a_5 \wedge a_4, a_6 \wedge a_5 \wedge a_4 \wedge a_2 \wedge a_0\} \tag{10}$$

where the concatenation operator "{,}" combines the bits of two or more data objects. In Eqs. (9) and (10), δ and Iδ are implemented by "XOR" gates denoted as "^" hereafter.

Likewise, the logic designs of A and IA can be represented respectively as

$$b = \{a_7 \wedge a_6 \wedge a_5 \wedge a_4 \wedge a_3, \sim a_6 \wedge a_5 \wedge a_4 \wedge a_3 \wedge a_2, \sim a_5 \wedge a_4 \wedge a_3 \wedge a_2 \wedge a_1, \\ a_4 \wedge a_3 \wedge a_2 \wedge a_1 \wedge a_0, a_7 \wedge a_3 \wedge a_2 \wedge a_1 \wedge a_0, a_7 \wedge a_6 \wedge a_2 \wedge a_1 \wedge a_0, \sim a_7 \wedge a_6 \wedge a_5 \wedge a_1 \\ \sim a_7 \wedge a_6 \wedge a_5 \wedge a_4 \wedge a_0\} \tag{11}$$

$$b = \{a_6 \wedge a_4 \wedge a_1, a_5 \wedge a_3 \wedge a_0, a_7 \wedge a_4 \wedge a_2, a_6 \wedge a_3 \wedge a_1, \\ a_5 \wedge a_2 \wedge a_0, \sim a_7 \wedge a_4 \wedge a_1, a_6 \wedge a_3 \wedge a_0, \sim a_7 \wedge a_5 \wedge a_2\} \tag{12}$$

where the "$\sim$" operator indicates a bit-wise logic inversion of each input bit. In such a way, the multiplication with constant λ and squaring in GF(2 ^ 4) are combined together to reduce the combinational logic and shorten the critical path as below:

$$\begin{aligned} b_3 &= a_2 {}^\wedge a_1 {}^\wedge a_0 \\ b_2 &= a_3 {}^\wedge a_0 \\ b_1 &= a_3 \\ b_0 &= a_3 {}^\wedge a_2 \end{aligned} \tag{13}$$

Similarly, the multiplication and the inversion in GF(2 ^ 4) in sub-stage1 can be implemented involving only "XOR" and "AND" gates.

In sub-stage2, the order of $I\delta \times A/I\delta$ and SR/ISR is exchanged, and then combined with the word-size input of the MC/IMC transformation. During the final adding round key transformation, a key is added to the state by a simple bitwise "XOR" operation. For the encryption engine, the 10-round keys from RK(0) to RK(a) are forward input, otherwise the direction is reversed from RK(a) to RK(0) for the decryption engine.

4 Experimental Results

In this section, the system performance metrics, involving area cost, power consumption, and transfer latency, are estimated.

4.1 Design Structure

Before discussing the detail implementation, the system performance is statically analyzed. Considering the speed requirement and the target device used in our work, the design guidelines are as follows.

(1) For the sequential part, less than 5 adders/subtractors (ADD/SUB) or only 1 multiplier (MUL) can be executed in one clock cycle. It avoids the setup timing violation and reduces the critical path.
(2) For the combinational part, we split the structure into subsections with approximately equal critical path delay so that the maximum throughput can be achieved.

Following these two guidelines, the calorie computation module is split into two cycles, and each round of the AES encryption/decryption is separated into two subsections. The design structure is shown in Fig. 4, and the resource cost and critical paths are summarized in Table 2.

For the encryption engine, the critical path of sub-stage1 has 15 "XOR" and 1 "MUX", and the critical path of sub-stage2 has 8 "XOR" and 1 "MUX". For the decryption engine, the critical path of sub-stage1 has 16 "XOR" and 1 "MUX", and the critical path of sub-stage2 has 11 "XOR" and 1 "MUX". Additionally the first cycle operates 4 "ADD" and the second cycle operates 1 "MUL" on the critical path.

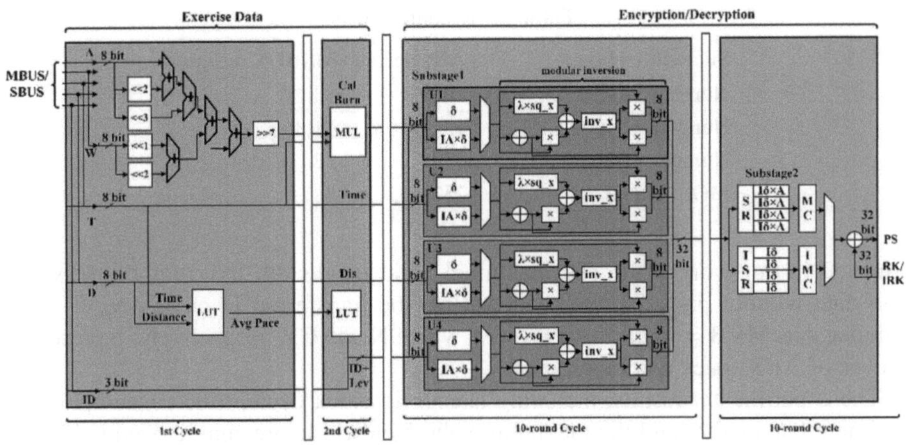

Fig. 4. Calculation and AES core design structure.

Table 2. Total logic and critical path

Cycles		Total logic	Critical path
1		5ADD + 1DIV	4ADD/1DIV
2		1MUL + 1MUX	1MUL
3–21	ENV	68XOR + 36AND + 1MUX	15XOR + 1MUX
4–22		128XOR + 1MUX	8XOR + 1MUX
3–21	DEC	75XOR + 36AND + 1MUX	16XOR + 1MUX
4–22		211XOR + 1MUX	11XOR + 1MUX

4.2 Area Cost and the Maximum Clock Frequency

In what follows, we design the circuit using Verilog hardware description language (HDL) and simulate the design-under-test (DUT) using Mentor Graphic ModelSim. As a case study, a direct case with 40-matrix training data, involving calorie calculation, memory write with decryption, and memory read with encryption, is tested. The transfer latency is run-time collected during the simulation.

The FPGA slice cost and the maximum operational frequency (MOF) are obtained by synthesis. In our work, we use Xilinx ISE 14.7 with Virtex5-110t as the target device. As the results summarized in Table 3, it can be observed that the MSBUS based design costs less IOs and slice count, and achieves higher maximum operation frequency, due to the compact structure.

4.3 Speed and Power

As full-duplex bus structures such as AXI and MSBUS, the maximum bandwidth (MB) can be directly formulated as MB = 2 × 4 × MOF unit in MBps. To concentrate on the transfer efficiency, the maximum valid bandwidth (MVB) is further

Table 3. Resource costs

Slice utilization	AXI design	MSBUS design
Number of slice registers	10,765	8,827
Number of slice LUTs	13,294	11,352
Number of bonded IOs	533	324
MOF (MHz)	252	287

estimated in our work [16]. The maximum valid bandwidth indicates the maximum valid data without protocol overhead that can be transferred in one cycle. It can be formulated as MVB = (MB $\times$ 4 N)/(CY), where N and CY, respectively, represent the number of AES states and clock cycles.

The experimental results, including latency in clock cycle, time consumption, the maximum bandwidth, and the maximum valid bandwidth, are summarized in Table 4. As a high-efficiency bus architecture, MSBUS design consumes less cycles than AXI for all the linear test, block test, and cipher test. To process 40-matrix or 160 users' training data, the MSBUS circuit spends 460 ns and the AXI bus consumes 683 ns in the cipher test.

Table 4. Transfer latency and valid bandwidth

Performance		Linear test	Block test	Cipher test
Clock cycles	AXI	92	98	172
	MSBUS	82	82	132
Latency (ns)	AXI	365	389	683
	MSBUS	286	286	460
Maximum bandwidth (Gbps)	AXI	2.02		
	MSBUS	2.30		
Maximum valid bandwdith (Gbps)	AXI	1.75	1.65	0.94
	MSBUS	2.24	2.24	1.39

Since the low-cost MSBUS structure achieves higher operational frequency and shorter latency, the maximum bandwidth of MSBUS is higher than AXI implementation. Considering the valid throughput, the number of valid data transferred by MSBUS per cycle can reach 1.39 GB per second, which is 1.5 times compared with AXI.

Furthermore, inputting the synthesis files (.NCD and .PCF) as well as the specific simulation file (VCD) into the XPower Analyzer tool, the power statistics are obtained in Table 5. Since static power consumption is mostly determined at the circuit level, it is almost a constant for different test cases as shown in the second and third rows in Table 5.

In the fourth and fifth rows, it can be observed that the MSBUS circuit consumes less dynamic power than AXI for all the linear, block, and cipher cases. For example, the dynamic power consumed by MSBUS is reduced to 74.4% compared with AXI for the cipher tests.

Table 5. Power consumption

Performance		Linear test	Block test	Cipher test
Static power	AXI	1398	1401	1403
	MSBUS	1377	1377	1378
Dynamic power	AXI	512	585	831
	MSBUS	488	501	618
Total	AXI	2010	2086	2214
	MSBUS	1965	1978	2096

5 Conclusions

This paper proposes a specific biofeedback system in order to instruct exercises by run-time measuring the heart rate and providing important training data, such as calorie consumption and the maximum aerobic training level. By adopting a cost-effective hardware structure, we not only realize the system functions, but also leverage the algorithm complexity and the limited resource on embedded systems.

In the future, we plan to provide more test vectors with external interfaces, such as directly regulating the exercise intensity using I2C or UART controllers and transmitting/receiving data to/from wireless hosts, to satisfy more applications for guaranteeing low-cost and low-power consumption.

References

1. Maffetone, P.: The Maffetone Method: The Holistic, Low-Stress, No-Pain Way to Exceptional Fitness. Ragged Mountain Press, Camden (2000)
2. Maffetone, P.: The Big Book of Endurance Training and Racing. Skyhorse Publishing, New York (2010)
3. Scalzi, S., Tomei, P., Verrelli, C.: Nonlinear control techniques for the heart rate regulation in treadmill exercises. IEEE Trans. Biomed. Eng. **59**(3), 599–603 (2012)
4. Schindelholz, M., Hunt, K.: Feedback control of oxygen uptake profiles during robotics-assisted treadmill exercise. IET Control Theor. Appl. **9**(9), 1433–1443 (2015)
5. Keytel, L.R., Goedecke, J.H., Noakes, T.D., Hiiloskorpi, H., Laukkanen, R., Van Der, M.L., Lambert, E.V.: Prediction of energy expenditure from heart rate monitoring during submaximal exercise. J. Sports Sci. **23**, 289–297 (2005)
6. Fan, M., Han, Q., Yang, X.: Energy minimization for on-line real-time scheduling with reliability awareness. J. Syst. Softw. **127**, 168–176 (2017)
7. AMBA Specification, Axis. Sunnyvale, CA, USA (1999)
8. AMBA AXI Protocol Specification, Axis. Sunnyvale, CA, USA, October 2011
9. Wishbone BUS, Silicore Corporation, Corcoran, MN, USA (2003)
10. Open Core Protocol Specification: OCP International Partnership, Beaverton, OR, USA (2001)
11. Yang, X., Andrian, J.: A high performance on-chip bus (MSBUS) design and verification. IEEE Trans. Very Large Scale Integr. (VLSI) Syst. **23**(7), 1350–1354 (2014)

12. Yang, X., Wen, W.: Design of a pre-scheduled data bus for advanced encryption standard encrypted system-on-chips. In: 2017 22nd Asia and South Pacific Design Automation Conference (ASP-DAC), January 2017
13. Advanced Encryption Standard (AES): FIPS-197, National Institute of Standards and Technology, November 2001
14. ANSI and IEEE: IEEE Standard for Binary Floating-Point Arithmetic. ANSI/IEEE Standard, Std 754-1985 (1985)
15. Kermani, M., Masoleh, A.: Design of A Pre-Scheduled Data Bus (DBUS) for Advanced Encryption Standard (AES) Encrypted System-on-Chips (SoCs). IEEE Trans. Comput. (TC) **61**(8), 1165–1178 (2012)
16. Yang, X., Wu, N., Andrian, J.: A novel bus transfer mode: block transfer and a performance evaluation methodology. Integr. VLSI J. **52**, 23–33 (2016)

Blast Performance of Demining Footwear: Numerical and Experimental Trials on Frangible Leg Model and Injury Modeling

Mehmet Karahan(✉) and Nevin Karahan

Vocational School of Technical Sciences, Uludag University, Gorukle, Bursa, Turkey
mkarahan@uludag.edu.tr

Abstract. This study reveals the protective efficacy of a personal protective boot against mine blasts both experimentally and numerically. By employing analyses conducted with the use of different amounts of explosives, the protective efficacy of the developed boot is compared to a typical military boot as reference. Both for analysis and verification tests, a ballistic gelatin covered frangible leg model was used. Strain energy that is exerted on the leg was determined with numerical analyses and verified with data obtained from strain gauges that were placed on the leg model. By employing the dynamic finite elements method, the physical injury that occurred to the leg model was examined and compared with the results of the blast test. The type-2 boot decreased the strain energy by approximately 80% compared with the type-1 boot. This observation was also verified with measurements obtained from strain gauge sensors placed along the tibia bone. It was observed that the damage occurring on the tibia bone was limited to local injuries and concentrated at a single spot without causing any fractures with the type-2 boot. As a result of axial load, the leg with the type-1 boot demonstrated fractures at several points rather than a fracture at a single point based on deflection. Hence, the latter case yields irremediable injuries.

Keywords: AP mine boot · Frangible leg model · FE modeling · Injury modeling

1 Introduction

Anti-personnel (AP) landmines are a global problem that affects many countries. Despite the recent attempts to prohibit the usage of these weapons, there are still a large number of landmines in the field and stockpiled that pose a constant threat to soldiers and civilians alike.

There are a few successful applications of anti-mine boots available. The *"Spider Boot"*, which was developed to protect against AP mines, is a commercialized design [1, 2]. The boots known as *"Over Boots"* are worn over a normal or protective boot. This design keeps the feet above the ground due to the thickness of the boot sole [3, 4]. Similarly, they are not suitable for long-term use due to the lack of mobility.

T. Ahram (ed.), *Advances in Human Factors in Sports, Injury Prevention and Outdoor Recreation*, Advances in Intelligent Systems and Computing 603,
DOI 10.1007/978-3-319-60822-8_6

Recent studies of protective boots have generally focused on designs that are similar to conventional army personnel boots. The early designs were reinforced with metal plates in the soles [5, 6]. Later studies developed lighter constructions by using Kevlar and Dyneema fabric layers with light metal alloys [7, 8]. Those built without any metal reinforcement had multiple layers of Kevlar and Dyneema composite plates [9]. One of the commercially successful models developed by Zeman Company included Kevlar layers in the sole that provide protection against a maximum of 50 g of explosive AP mines [10, 11]. The latest designs have an energy absorbing material under the composite layers to reduce the magnitude of the incident wave simply by inelastic deformation [12–14]. Such components are generally in the form of V-shaped deflectors in the sole to reflect the incident blast wave, thus reducing the momentum transmitted to the feet [15].

A common characteristic of protective boots or footwear is the polymer composite materials in the soles to protect against the blast effects. However, the use of only composite layers cannot provide sufficient energy absorption for use against AP landmines. Using plastically deformable and compressible foam and honeycomb materials with composite layers considerably increases energy absorption [16].

A frangible leg with a geometry corresponding to the 50th percentile of Australian males has been developed [17]. This leg includes all of the major bones, which are cast using a synthetic material. The bones are then assembled with adhesives and simulated tendon materials. The resulting structure is then placed in a larger mold corresponding to the outer shape of the human leg, and gelatin is cast around the structure to simulate the soft tissue [17, 18].

Although there are some experimental studies to evaluate the effectiveness of the protection of AP mine boots, in this field, comprehensive and verified finite element analysis is not available.

This study is part of a research series. The protection efficiency of different protective boots has been researched with a mechanical [19] and frangible [20] leg in previous studies. This study focuses on the protection effectiveness of a personal protective boot against mine blasts, both experimentally and numerically. By employing analyses that are conducted with the use of different amounts of explosives, protective efficacy of the developed boot is evaluated by using a typical military boot as reference. For analysis and verification tests, a ballistic gelatin covered frangible leg model was used.

2 Experimental and Numerical Study

In this study, two types of boots were used. One type is a standard military boot (Type-1). It was used as a reference to compare the protection levels of the other boot. There are no protective insert layers in the sole and shoe upper of this boot. The other type is a protective boot with reinforced soles (Type-2).

The general appearance of the protective boot used in the tests is shown in Fig. 1. The boot sole consists of two layers. The first layer, which is the bottom layer, has a deflector to redirect the shockwave. The second layer, which is on top of the deflector, consists of composite inserted plates. The shoe upper of the boot is strengthened using a two-ply aramid-woven fabric between the lining and the leather.

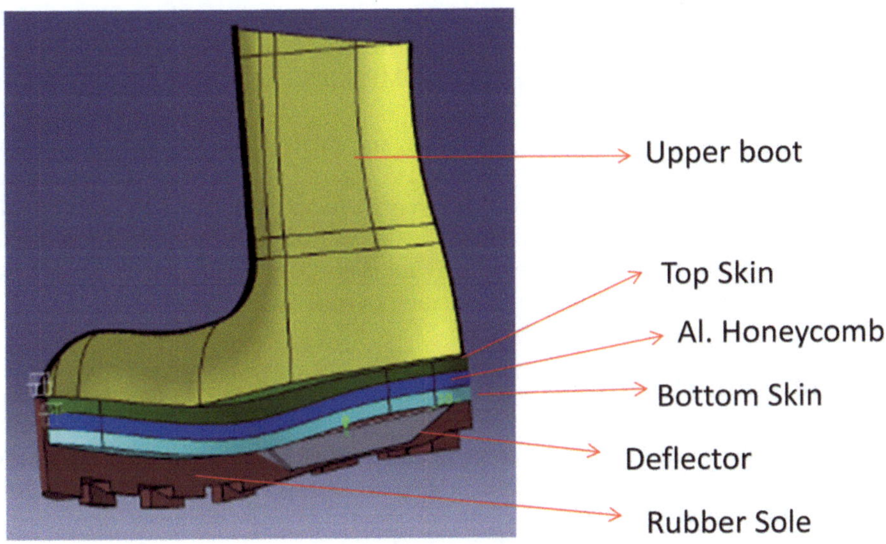

Fig. 1. General view of protective boot sample.

The composite plate using in the sole was produced with aramid Uni-Directional (UD) fabric layers and Nolax A21.2007 low-density polyethylene (LDPE) adhesive film. Aluminum honeycomb with 3 mm cells was inserted between two layers of composite skins with glue.

The frangible leg model (FLM) for the second test series used tibia and fibula bones that are virtually identical to the 42nd percentile human leg based on bone size and geometry. The bones of the legs were manufactured using poly(methyl methacrylate) (PMMA) and constructed around a porous core so that the overall static strengths of these soft bones were comparable to the strengths of their human counterparts. Strain gauges were attached along the tibia with a tape. Wires used for data gathering were protected in hoses. An 8-channel LTT data acquisition system was used for the experiments.

The FLM was also fitted with adaptor plates to mount the test frames. These plates were fixed to the top of the tibia using nylon-threaded inserts and nuts. The upper ends were drilled and tapped to fit an eyebolt so that the finished FLM could be attached to the crossbeam above the test platform.

The next step was to add gelatin to completely cover the FLM. Nylon skin was then fitted over the FLM in four layers.

After the FLM was placed in a boot, an explosive was buried at a depth of 70 mm in the soil with the center of the heel barely touching it, which is considered to be the worst case. The soil, which was placed inside plastic containers for these tests, was medium sand that was purchased locally in bags. 40 g C4 explosives were used in the tests.

Two types of boots were used for finite element (FE) analyses. Among these, Type-1 is a classical military footwear and it does not have a deflector or additional composite layer on its sole. However, Type-2 boot is reinforced with deflector and sandwich composite layers in the sole. Input data of composite layers located on the top of the deflector in the Type-2 sample has been created as a results of test series [21–23]. ANSYS LS-Dyna-based Orthotropic MAT COMPOSITE FAILURE SOLID MODEL (59) material cards used.

Employing the model, FE method were run by separately simulating a squeeze in three directions and force-stroke curves were obtained in 3 axes. These curves were transformed into axial stress-volumetric strain curves and were included in the MAT HONEYCOMB (26) material model. Rate dependent plasticity model of outer skin steel material of deflector was dissolved using empirical Johnson-Cook material model (MAT JOHNSON COOK (15)) and Grüneisen equation of state (EOS_GRUNEIS).

3 Results and Discussions

For the case where there is no protection against a mine blast, 40 g of C4 explosives was detonated at the heel. Fracture of the tibia bone after 2 ms is shown in Fig. 2 with further details. Experimental observations are quite similar to the result illustrated in Fig. 2. In addition to the tibia bone being broken at several points, fragments were shattered from the blast point. Thus, the load directly touched the system. However, since the explosive was buried under the soil, there was no burning damage, as opposed to the previous study [20]. In any case, bone fractures are quite serious, as observed in the previous experiments. Since the boot completely ruptured, the leg was directly exposed to dust and particles. Therefore, irremediable open injuries are highly possible.

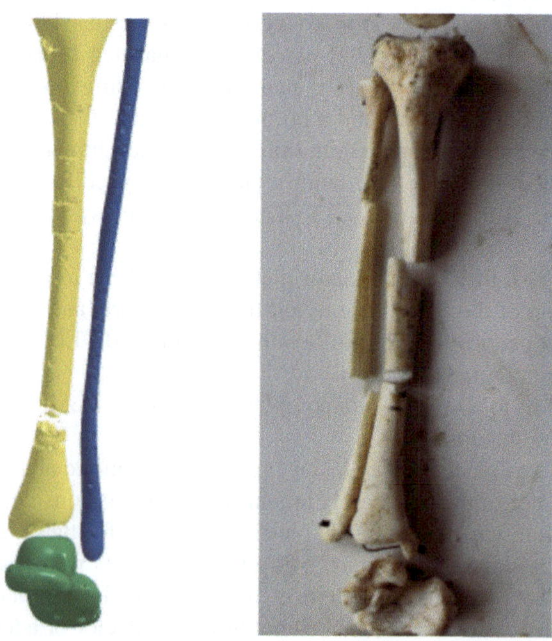

Fig. 2. Numerical and experimental results of frangible leg worn Type-1 boot sample after explosion.

The time-dependent behavior of the accumulated strain energy at the tibia bone is shown in Fig. 3. Maximum strain energy is 98.19 J at 0.32 ms.

Deformations occurring at specific time intervals are shown in Fig. 4. As observed in Fig. 4, deformation caused by the blast compresses the deflector upwards, while the honeycomb between the composite layers is partially compressed and deformed. Being buried under the soil, the impulse created was reduced, even though the explosive was strong. Experimental results also confirmed this situation. The experimental and numerical results are shown comparatively in Fig. 4. According to experimental results, even though the deflector was completely deformed, the honeycomb core was partially compressed and deformed.

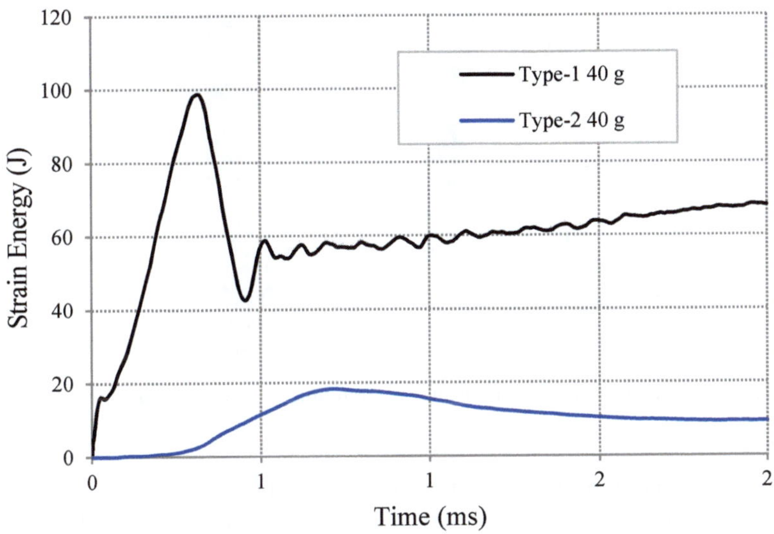

Fig. 3. Calculated strain energy at the tibia bone.

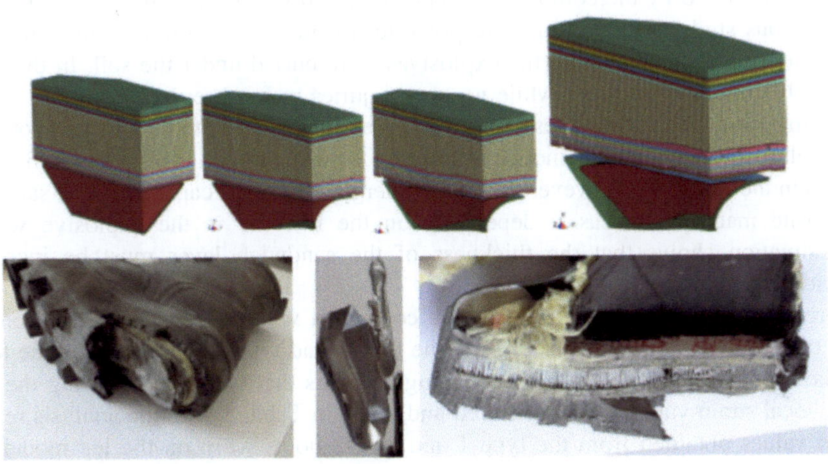

Fig. 4. Numerical and experimental evaluation of damage of Type-2 sample after explosion.

For detail analysis for Type-2 sample, the load-time plot obtained at the first step was applied to the sole with 40 g of C4 explosives. This way, lower leg injury was estimated for the lower leg model with the type-2 boot. The analysis was executed for 2.0 ms, until the deformation on the tibia bone reached a steady state. Deformations on the lower leg at specific time points as well as fractures on the tibia bone are shown in Fig. 5.

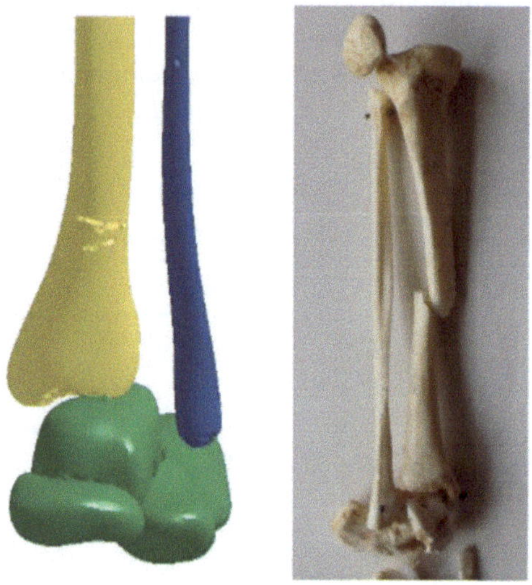

Fig. 5. Numerical and experimental results of frangible leg worn Type-2 boot sample after explosion.

Results obtained from the experimental test are quite similar to the results shown in Fig. 5. Only a single fracture was observed on the tibia bone. Additionally, the ballistic gel representing the soft tissues and broken pieces were trapped inside the boot and did not burst from the blast point. This situation indicates that a large portion of the blast load was absorbed by the composite layers. Observed injuries are milder than those of the previous study. Although a more powerful explosive was used in this study, the blast load was decreased since the explosives were buried under the soil. In this case, remediable injuries occurred, while no open injuries were observed.

Using sandwich composites with a compressible core material on the protective boot sole provides a better energy absorption compared with only monolithic composites in the sole [19]. However, the strain energy absorption capacity of the sandwich composite material decreases, depending on the increase in the explosive weight. This situation shows that the thickness of the sandwich layer must be increased depending on the explosive weight.

From these results, FE analyses were confirmed with the data acquired from the blast tests, which were conducted using the type-1 and type-2 boot worn on the lower leg model. With the help of the strain gauge sensors that were placed along the tibia bone, local strain values were measured and given in Table 1 with the analysis results. Strain values obtained from the type-1 and type-2 boots worn on the leg models are

Table 1. Comparison of data obtained from experimental study and FE analysis results

Tibia strain energy values obtained with FE analysis (J)		Rate (a/b)	Strain values obtained from experimental study		Rate (c/d)	Differences (%)
Type-1 (a)	Type-2 (b)		Type-1 (c)	Type-2 (d)		
98,19	18,48	0,188	0,040	0,013	0,325	**0,138**

similar to the analysis results, and the percentage difference between the two values was limited within ±0.15. Therefore, numerical analyses and experimental results show correlation.

4 Conclusions

This study includes the results of experimental and numerical tests performed using a frangible leg model to compare the protection efficiencies of different boots. The applied test method was compatible with previous studies [18, 24]. The results that were achieved are as follows:

- Using sandwich composites with a compressible core material on the protective boot soles provides better energy absorption. In the type-2 boot samples, which had soles with this form, the strain energy was reduced approximately 80% compared to the type-1 boot. This observation was also confirmed by local measurements conducted via strain gauge sensors that were placed along the tibia bone.
- The energy absorption capacity of the sandwich composite material decreases, depending on the increase in the explosive weight. This situation shows that the thickness of the sandwich layer must be increased depending on the explosive weight.
- Using sandwich composites with a compressible core material on the protective boot sole definitely provides a better energy absorption compared with only monolithic composites in the sole [19].
- The qualitative observation of the damage to the tibia bone suggests that the leg used for the type-2 boot has local damage that occurred at a single point without any rupture.
- It was determined that the deformation observed on the articular surface was due to the removal of the cartilage tissue elements reaching the strength criterion. The damage observed in the lower legs with the type-1 boot was due to fragmentation at many points as a result of axial loading rather than a single-point bending fracture. These findings have been confirmed experimentally.

References

1. Shariful, I., Makris, A., Bergeron, D.: The Spider Boot: an effective foot protective systems against anti-personnel mine blasts. J. Conventional Weapons Destruction **4**(2), 1–7 (2000)
2. Makris, A., L'Abbe, R.J., Voisine P., et al.: Anti-personnel mine foot protection systems. US patent no: 6.006.646 (1999)

3. Krohn, H.F., Flynn, R.M., Sasaman, W.O.: Protective footpad assembly. US patent no. 2.720.714, 15 August 1952
4. Chavet, I., Madmoni, A.: Minefield shoe and method for manufacture thereof, US2003/0172554 A1, 18 September 2003
5. Ringler, S., Chavet, I.: Device for reducing the danger of accidental detonation of a landmine, US patent no: 4.611.411, 16 September 1986
6. Jordan, R.D.: Protective footgear, US patent no. 3.516.181, 23 June 1970
7. Fujinaka, E.S., et al.: Blast protective footwear, US patent no: 3.318.024, 31 May 1996
8. Vaz, G.A.: Protective boot and sole structure, US 2002/011011 A1, 31 January 2002
9. Vaz, G.A.: Blast and fragment resistant safety boot footwear, US patent no: 5.979.081, 9 November 1999
10. Joynt, V.P., Dyk, J.T.V.: Protective footwear, US 2006/0000117 A1, 5 January 2006
11. Zeman, P.: Anti land mine-boots, WO 03/037125 A1, 29 October 2001
12. Lohrmann, H.R.: Anti-personnel mine protective footpad, US patent no: 5.992.056, 30 November 1999
13. Peche, J.P., Gaultier, R., Peltzer, M.O.: Appliance for protecting against the effects of explosive devices, US patent no: 6.655.051, 2 December 2003
14. Kamberoglu, M., Karahan, M., Alpdogan, C., Karahan, N.: Evaluation of foot protection against AP mine blast: effect of deflector geometry. J. Test. Eval. (2016). doi:10.1520/JTE20150171
15. Trimble, K., Clasper, J.: Anti-personnel mine injury mechanism and medical management. J. Royal Army Med. Corps **147**, 73–79 (2001)
16. Hanssen, A.G., Enstock, L., Langseth, M.: Close-range blast loading of aluminum foam panels. Int. J. Impact Eng. **27**, 593–618 (2002)
17. Krstic, A.R.: Australian Patent No: AUPQ8943 entitled 'A Surrogate', July 2000
18. Bergeron, D.M., Coley, G.G., Fall, R.W., Anderson, I.B.: Assessment of Lower Leg Injury from Land Mine Blast – Phase 1: Test Results Using a Frangible Surrogate Leg with Assorted Protective Footwear and Comparison with Cadaver Test Data. DRDC Suffield TR 2006-051. Defence R&D, Suffield, Canada (2006)
19. Karahan, M., Karahan, E.A.: Development of an innovative sandwich composite material for protection of lower limb against landmine explosion: mechanical leg test results. Textile Res. J. (2016). doi:10.1177/0040517515624880
20. Karahan, M., Karahan, N.: Development of an innovative sandwich composite material for protection of lower limb against landmine explosion. J. Reinf. Plast. Compos. (2016). doi:10.1177/0731684416668261
21. Shaker, K., Jabbar, A., Karahan, M., Karahan, N., Nawab, Y.: Study of dynamic compressive behaviour of aramid and ultrahigh molecular weight polyethylene composites using Split Hopkinson Pressure Bar. J. Compos. Mater. (2016). doi:10.1177/0021998316635241
22. Karahan, M., Jabbar, A., Karahan, N.: Ballistic impact behavior of the aramid and ultra-high molecular weight polyethylene composites. J. Reinf. Plast. Compos. **34**(1), 37–48 (2015)
23. Karahan, M., Yıldırım, K.: Low velocity impact properties of aramid and UHMWPE composites. Fibres Textiles Eastern Eur. 23, 3(111), 97−105 (2015)
24. Bergeron, D.M., Coley, G.G., Fall, R.W., Anderson, I.B.: Assessment of Lower Leg Injury from Land Mine Blast – Phase 2: Follow up Tests with a Modified Frangible Surrogate Lower Leg and Comparison with Cadaver Test Data. DRDC Suffield TR 2007-045. Defence R&D, Suffield Canada (2007)

The Effects of Cupping Therapy on Reducing Fatigue of Upper Extremity Muscles—A Pilot Study

Chien-Liang Chen[1], Chi-Wen Lung[2], Yih-Kuen Jan[3,4,5], Ben-Yi Liau[6], and Jing-Shia Tang[7,8(✉)]

[1] Department of Physical Therapy, I-Shou University, Kaohsiung, Taiwan
chencl@isu.edu.tw
[2] Department of Creative Product Design, Asia University, Taichung, Taiwan
[3] Rehabilitation Engineering Laboratory, University of Illinois at Urbana-Champaign, Champaign, IL, USA
[4] Kinesiology and Community Health, University of Illinois at Urbana-Champaign, Champaign, IL, USA
[5] Computational Science and Engineering, University of Illinois at Urbana-Champaign, Champaign, IL, USA
[6] Biomedical Engineering, Hungkuang University, Taichung, Taiwan
[7] Department of Nursing, Chung-Hwa University of Medical Technology, Tainan, Taiwan
tangjeanine@yahoo.com.tw
[8] Institute of Allied Health Sciences, National Cheng Kung University, Tainan, Taiwan

Abstract. The aim of this study was to determine the effect of the cupping therapy on the muscles and autonomic nervous system. Five healthy participants were recruited for the arm cranking tests. The protocols include a 1st 12-min bout, 20-min recovery, and 2nd 12-min bout. The cupping therapy (negative pressure of 300–450 mmHg on the upper back and shoulder muscles) was applied during the recovery period of experimental trials. The median frequency (MDF) of electromyographic signals and the heart rate variability (HRV) were used to objectively quantify the degree of muscle fatigue and cardiovascular regulations. The repeated measures ANOVA was applied to determine the differences. Our results showed that MDF was significantly higher in the experimental trials but the HRV was not significantly different between the experimental and control trials. Our study indicates that the role of cupping therapy may reduce muscle fatigue but do not alter the cardiorespiratory controls.

Keywords: Dry cupping · Fatigue recovery · Electromyogram · Heart rate variability · Oxygen consumption · Rating of perceived exertion

T. Ahram (ed.), *Advances in Human Factors in Sports, Injury Prevention and Outdoor Recreation*, Advances in Intelligent Systems and Computing 603,
DOI 10.1007/978-3-319-60822-8_7

1 Introduction

The stressful components of training and competitions may temporarily impair an athlete's performance. An adequate recovery, however, decreases fatigue, accelerates the rate of physiological regeneration, may decrease the risk of injury, and enhances supercompensation [1]. Therefore, successfully answering the question of how to facilitate a rapid recovery of exercise performance is often considered to be the key factor in winning. Various recovery modalities, including sports massage and various water therapies, are currently used; but it is important to determine the effectiveness of and understand the rationale underlying each modality [2]. Recently, cupping therapy has become a popular postexercise recovery intervention used by athletes taking part in various sports and events; however, there is still not enough evidence to support the use of cupping to improve exercise performance.

Cupping is an ancient therapy that has most commonly been used in Middle Eastern and Asian countries. The placement of the cupping cup creates a partial vacuum, which is widely believed to stimulate muscles and blood flow in the region of the body upon which the cup is placed. Cupping is thought to mainly act by increasing local blood circulation and relieving painful muscle tension [3]. It is believed that it mainly involves improvements in microcirculation, promoting capillary endothelial cell repair in the regional tissues. This helps in normalizing the patient's functional state and progressive muscle relaxation [4]. A previous study has indicated that cupping is the best deep tissue massage [5]. It is obtained to stimulate muscles similar to other therapies like tapping, slapping or pummeling [6]. In addition, cupping therapy restores sympathovagal imbalances by stimulating the peripheral nervous system [7].

It has been proven that the recovery rate from muscle fatigue and autonomic nervous system activity after exercise is significantly correlated with the exercise capacity [8, 9]. It was therefore assumed that the appropriateness of cupping therapy in improving recovery from muscle fatigue would be an issue worth exploring. As such, this study sought to investigate whether 15 min of dry cupping after the inducement of exercise fatigue would speed up recovery, and, if so, to clearly define the physiological mechanism by which it does so.

2 Methods

2.1 Participants

A total of 5 moderately trained people (4 men and 1 woman) participated in this study. Their mean age, height, weight, and body mass index were 25.9 ± 9.3 years, 172.4 ± 4.8 cm, 68.3 ± 9.7 kg, and 22.9 ± 2.9 kg/m^2, respectively. All the participants were routinely involved in various intermittent sports activities (e.g., volleyball, basketball, and badminton), were familiar with maximal training, and had no history or clinical signs of cardiopulmonary diseases or orthopaedic injury in the upper extremities. The participants were fully informed about the study protocol and gave their informed written consent to participate in the experimental procedure. All applicable institutional and governmental regulations concerning the ethical use of human volunteers were followed during the course of this research.

2.2 Design

Each participant visited our laboratory three times to complete an arm cranking exercise. At the first visit, an incremental maximal exercise test was performed by the given participant in order to determine his or her peak oxygen consumption (VO_2peak). This result was then used to determine the exercise intensity that would cause exercise fatigue for that participant, as exercise fatigue would subsequently be induced in order to observe the effects of cupping on the physiological responses to fatigue. During the next two visits, a control trial and an experimental trial (which included the use of cupping) were performed in random order. Each exercise trial was conducted in an air-conditioned laboratory with an atmospheric temperature of 21–25 °C and a relative humidity of 50%–60%. Each participant completed the experimental protocol within a period of 14 days, with at least 3 days between each exercise test, to ensure that the participant's level of physical fitness had not changed over the course of the visits. In addition, all the participants were asked to undergo the 15 min of dry cupping alone before the first exercise test in order to determine the simple effect of cupping on heart rate variability (HRV).

2.3 Dry Cupping Intervention

The participants were seated in a leaning-forward position, using a table for upper extremity support, and cupping was performed as follows: Acrylic cups 5 cm in diameter were used. Six points of the bilateral deltoid, upper trapezius, and lateral area of the latissimus dorsi muscles were selected for treatment. Cups were placed on these points, and negative pressure was applied by a cupping pump and retained for 15 min. The negative pressure applied ranged from 300–450 mmHg and was adjusted according to the patient's tolerance [10].

2.4 Maximal Arm Cranking Exercise Test

The participants performed the incremental maximal exercise test by using an arm ergometer (ANGIO with a chair; Lode, Groningen, The Netherlands). The testing protocol was preceded by a 1-minute warm-up period at 0 W, after which the workload was increased by 15 W/min (ramp protocol) until the participant felt exhausted [11]. The pedaling rate was maintained at approximately 60 rpm for each participant. A pedal frequency meter with visual feedback was used by each participant to maintain the pedaling rate. The test was terminated when the given participant could not continue because of exhaustion or when the target pedal rate could not be maintained for 10 s despite verbal encouragement. All the participants achieved at least two of the following criteria for determining maximal effort: (1) a plateau in oxygen consumption

(VO_2) with an increased work rate; (2) a heart rate > 85% of the age-predicted maximum; or (3) a respiratory exchange ratio (RER) greater than 1.1 [12].

Expired air was analyzed breath-by-breath by using an automated system (Vmax 29c; Sensor Medics, Yorba Linda, CA, USA). The measured exercise cardiorespiratory parameters were the heart rate, VO_2, carbon dioxide production (VCO_2), minute ventilation (V_E), RER, and the ventilatory equivalents for O_2 (V_E/VO_2) and CO_2 (V_E/VCO_2). The V-slope and ventilatory equivalence methods were used to evaluate the ventilatory threshold (VT) in a combined model [12]. The V-slope method, which is the most common method, entails graphing VCO_2 versus VO_2 values. The VT is then identified as the point at which there is a shift in the slope along a line identified between these gas measurements. In addition, the ventilatory equivalence method is defined as the intensity of activity that causes the first rise in the V_E/VO_2 without a concurrent rise in the V_E/VCO_2. If the V-slope method cannot provide a reliable VT, it is common to review the V_E/VO_2 and V_E/VCO_2 curves to determine the VT.

2.5 Exercise Protocols

Participants came back at least 2 days after the maximal exercise test to perform two consecutive exercise bouts, each of which was 12 min in duration. The first exercise bout consisted of 6 min of moderate-intensity constant-load exercise and 6 min of high-intensity constant-load exercise. Before the exercise bout itself, however, there was a 1-minute, 0 W warm-up, and the bout was followed by a 2-minute, 0 W cool-down (Fig. 1). The moderate-intensity part of this exercise bout was performed at a power output (PO) corresponding to ~80% of the PO at VT; the high-intensity part was performed at a PO corresponding to ~50% between the PO at VT and the PO at VO_2peak. The pedaling rate was maintained at approximately 60 rpm throughout the course of the experiment. Following a 20-minute period of recovery, the participant performed a second exercise bout with the same content as the first exercise bout.

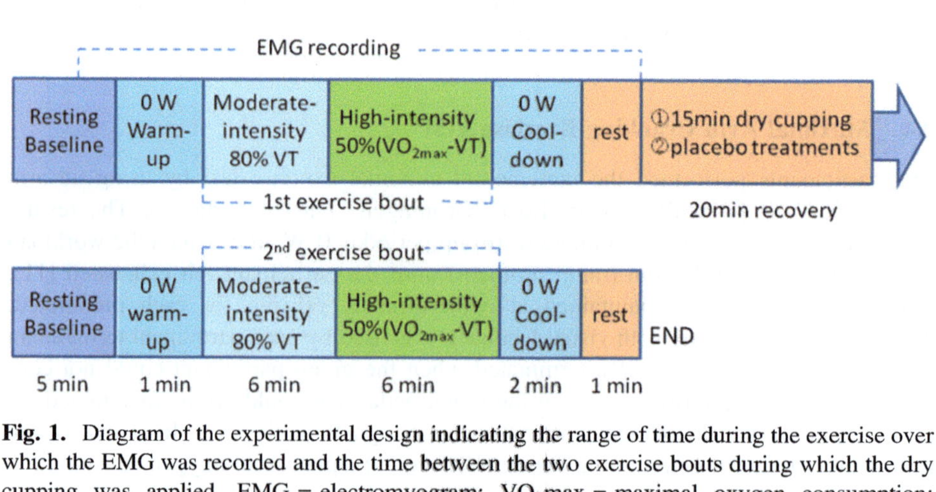

Fig. 1. Diagram of the experimental design indicating the range of time during the exercise over which the EMG was recorded and the time between the two exercise bouts during which the dry cupping was applied. EMG = electromyogram; VO_2max = maximal oxygen consumption; VT = ventilator threshold.

For the experimental trial, the participants were required to undergo dry cupping treatment with a negative pressure of 300–450 mmHg for 15 min during the 20-minute recovery period. For the control (placebo) trial, the participants instead only underwent cupping with a pressure of less than −100 mmHg; this low negative pressure was only used to ensure that the cups did not fall off. Furthermore, for both types of trial, the degree of subjective fatigue was assessed by the Borg's rating of perceived exertion (RPE) scale. After each high-intensity exercise bout, the subjects were immediately asked about their degree of exercise-induced fatigue in order to assess the impact of cupping on this aspect of self-perception during exercise.

2.6 Electromyography (EMG) Recordings and Analysis

Myoelectric activities were determined using surface EMG and recorded using bipolar silver-silver chloride electrodes that were 7 mm in diameter and fixed at a 20-mm interelectrode distance (Norotrode 20, Myotronics-Noromed, Inc., Tukwila, WA, USA). The EMG signals were amplified with a bandwidth of 20 to 450 Hz and a gain of 1000 (EMG 100C, Biopac Systems, Goleta, CA, USA), sampled using a data acquisition system (MP 150, Biopac Systems) at a rate of 1000 Hz, and subsequently stored on a computer disk for subsequent analysis. The electrodes were placed on the triceps brachii of the right arm, because these muscles were heavily involved in performing the arm cranking movement [13]. Each electrode site was prepared by abrading and swabbing the site by using an ether pad. The EMG signals were examined for movement artefacts, and the electrodes were secured using surgical tape to minimise displacement during movement. To ensure that the electrodes were placed at precisely the same location for each testing session, the electrode sites were marked using a pen. EMG signals were recorded continually during the two exercise bouts; the raw EMG signals were integrated using commercially available software (MATLAB; MathWorks Inc., Natick, MA, USA). Myoelectric signals were collected during the 10 s at the end of the last minute of high-intensity exercise. The standard periodogram method based on the short time Fourier transform was used to estimate the power spectrum of surface EMG, and the median frequency of the power spectrum (MDF) was then calculated.

2.7 Measurement and Analysis of the HRV

HRV is the physiological phenomenon of variation in the time interval between heartbeats. It is measured by the variation in the R–R intervals [the intervals between R waves on the electrocardiogram (ECG)]. Electrophysiological signals were recorded for each participant by using a miniature physiological signal recorder (TD1; Taiwan Telemedicine Device Company, Kaohsiung, Taiwan). In order to determine the simple effect of dry cupping on HRV, the HRV of each participant was measured before the participant performed the first exercise test. Specifically, the HRV was measured for a 5-minute period beginning 10 min before the participant received the cupping treatment and then for another 5-minute period beginning immediately after the participant received 10 min of the cupping treatment. During the exercise procedure, we compared the 20-minute HRV recovery values for the cupping and placebo trials (Fig. 1). The time points for this analysis consisted of three 5-minute periods: (1) The last 5 min of the high-intensity exercise; (2) The 10th minute of the recovery period after the first bout of exercise ended; and (3) The last 5 min of the 20-minute recovery period. During the assessment period, participants were monitored using ECG leads to record their heart rate and R–R intervals. The signals were recorded in real time after analog-to-digital conversion (8-bit) at a sampling rate of 500 Hz. The R–R intervals (in milliseconds) were calculated beat-to-beat by using a customized software program developed by Dr. Terry B. J. Kuo [14]. Frequency domain analysis was performed by nonparametric fast Fourier transform. The power spectrum was then quantified into frequency domain measurements. Total power (0–0.4 Hz), low-frequency (LF, 0.04–0.15 Hz), and high-frequency (HF, 0.15–0.4 Hz) power components were converted into absolute values of power, and any skewness in the distribution of the parameters was corrected by transforming all of them logarithmically. Total power is a marker of autonomic nervous activity, HF power reflects parasympathetic nervous activity, and LF power reflects partial contributions from both sympathetic and parasympathetic nervous activity. To detect the sympathetic effect on HRV, total power was used to normalize LF or HF power, which was expressed as a percentage (LF% or HF%). The ratio of low to high frequencies (LF/HF) was also calculated and used as an index of the sympathovagal balance [15].

2.8 Statistical Analysis

Data are presented as the mean and standard error (mean ± SD). Statistical differences in the changes of HRV parameters after cupping alone and the percentage changes of EMG MDF on the triceps between cupping and control trials were calculated using paired-sample *t* tests. A two-way (trial × time) repeated-measures analysis of variance (RM-ANOVA) was used to analyse the dependent variables over time in order to compare those values for the cupping and control trials. For the high-intensity portion of each bout of exercise during both trials, the RPE, MDF of triceps, and cardiorespiratory values were calculated for each evaluation. The analyses were performed using the data from the first and second exercise bouts to determine the difference in the fatigue responses between the cupping and control trials. In the recovery period between the two exercise bouts, the analyses were performed using the HRV data from the last 5 min of high-intensity exercise, the 10th min of the recovery period, and the end of the 20-minute recovery period of the two trials. This was done to determine whether cupping affects the recovery rate of autonomic nervous activity. Statistical analyses were conducted using PASW Statistics 18.0 (SPSS Inc., Chicago, IL, USA) software. The significance was set at $p < 0.05$, and trends were noted for $0.05 < p < 0.1$.

3 Results

The RPE scores during the two exercise bouts were compared for the cupping and control trials. The RM-ANOVA indicated a trend of trial-by-time interaction on the RPE ($p = 0.089$) and a significant main effect of trials ($p = 0.025$), but no main effect of time (Table 1). It was shown that the feeling of exertion when completing the second exercise bout was slightly more unpleasant than the feeling of exertion when completing the first bout in the control group. However, the participants in the cupping group reported greater consistency in their feelings in completing these two exercises bouts.

Table 1. Comparison of RPE scales at exercise bouts before and after cupping, and the interaction effect according to RM-ANOVA (n = 5) (mean ± SD)

Borg's RPE 6–20 scale	1st exercise	2nd exercise	P (F) value		
			Trial	Time	Trial × time
Control trial	16.6 ± 1.36	17.2 ± 1.72	0.025**	0.778	0.089*
Cupping trial	16.4 ± 1.36	16.0 ± 1.79	(12.250)	(0.091)	(5.000)

RPE = Rate of perceptive exertion; RM-ANOVA = repeated measures analysis of variance.
*Trend for significant difference ($0.05 < p < 0.1$);
**Significant difference ($p < 0.05$)

The EMG MDF of triceps was analyzed for the two exercise bouts in order to compare between the cupping and control trials. The RM-ANOVA indicated an obvious trial-by-time interaction on the MDF ($p = 0.032$), but no main effects of time or trial (Table 2). In order to express relative muscle activation, the EMG data were normalized. The value of MDF decreased by 4.27 ± 0.07% in the second exercise bout compared with the first bout in the control trial. However, the value of MDF increased by 8.88 ± 0.08% in the second exercise bout compared with the first bout in the cupping trial. There was a significant difference ($p = 0.033$) in the MDF change ratio between the cupping and control trials. These data show that lower fatigue levels were exhibited in the exercise responses to the cupping trial than in those to the control trial.

Table 2. Comparison of median frequency of the triceps EMG during high-intensity exercise before and after cupping, and the interaction effect (n = 5) (mean ± SD)

MDF (Hz)	1st exercise	2nd exercise	P (F) value		
			Trial	Time	Trial × time
Control trial	82.8 ± 4.9	79.2 ± 7.2	0.812	0.562	0.032**
Cupping trial	78.6 ± 9.7	85.4 ± 10.7	(0.065)	(0.399)	(10.440)

EMG = electromyogram; MDF = median frequency
*Trend for significant difference ($0.05 < p < 0.1$);
**Significant difference ($p < 0.05$)

The RM-ANOVA comparison of the values of heart rate, VO_2, VCO_2, and the minute ventilation collected during the two exercise bouts indicated that these values were similar for the cupping and control trials ($p > 0.05$), and no significant trial-by-time interaction was observed ($p > 0.05$) (Table 3).

Table 3. Comparison of cardiorespiratory responses during two exercise bouts before and after cupping, and the interaction effect (n = 5) (mean ± SD)

Control trial			Cupping trial		P value		
Parameters	1st exercise	2nd exercise	1st exercise	2nd exercise	Trial	Time	Trial × time
Heart rate (bpm)	153.6 ± 20.0	154.6 ± 23.8	149.8 ± 25.1	150.2 ± 20.1	0.616	0.800	0.883
VO_2 (L/min)	1.54 ± 0.36	1.46 ± 0.36	1.47 ± 0.38	1.36 ± 0.42	0.242	0.190	0.880
VCO_2 (L/min)	1.84 ± 0.41	1.75 ± 0.47	1.81 ± 0.53	1.72 ± 0.55	0.827	0.319	0.996
Ventilation (L/min)	52.2 ± 14.1	49.5 ± 15.2	50.9 ± 18.2	47.0 ± 18.1	0.66	0.135	0.893

VO_2 = oxygen consumption; VCO_2 = carbon dioxide production

To determine the effect of cupping on the autonomic nervous system at rest, we observed immediate HRV changes after the participants received dry cupping and before they started the exercise test. Both the mean R–R interval and HF showed an increased tendency after the application of cupping ($p = 0.06$) (Table 4). However, the total power, LF, LF/HF ratio, HF%, and LF% did not change significantly.

Table 4. Immediate effects of dry cupping on heart rate variability (n = 5)

Heart rate variability	Before cupping (mean ± SD)	After cupping (mean ± SD)	P value
RR (ms)	860.98 ± 105.25	912.09 ± 106.26	0.063*
TP [$\ln(ms^2)$]	7.77 ± 0.67	7.82 ± 0.98	0.816
LF [$\ln(ms^2)$]	6.97 ± 0.75	6.97 ± 1.19	0.985
HF [$\ln(ms^2)$]	5.84 ± 1.00	6.08 ± 1.00	0.061*
LF/HF	1.12 ± 0.63	0.90 ± 0.90	0.405
LF (%)	70.02 ± 11.33	67.71 ± 18.35	0.637
HF (%)	18.61 ± 7.86	23.22 ± 16.08	0. 441

RR = mean of R-R intervals; TP = Total power, marker of autonomic nervous activity; LF = low frequency power reflects both sympathetic and parasympathetic modulations; HF = high frequency reflects parasympathetic activity; LF/HF (ln ratio) = the ratio of LF to HF reflects sympathovagal balance; LF% = LF in normalized unit reflects sympathetic activity; HF% = HF in normalized unit reflects sympathetic inhibition.
*Trend for significant difference ($0.05 < p < 0.1$);
**Significant difference ($p < 0.05$).

Furthermore, we investigated the recovery rate of HRV during the 20-minute recovery period after the first bout of exercise in order to compare the HRV results for the cupping and control trials. Time exerted a significant main effect on most HRV variables during the recovery period ($p < 0.05$), except the changes in the ratio of LF/HF (Table 5). This showed that the variables at the end of the first exercise bout

Table 5. Comparison of the HRV recovery between control and cupping trial after higher-intensity exercise (n = 5) (mean ± SD)

HRV	Control trial			Cupping trial			P value		
	End of exercise	10 min recovery	20 min recovery	End of exercise	10 min recovery	20 min recovery	Trial	Time	Trial x time
RR	415.9 + 52.0	721.7 + 134.3	722.4 + 127.7	410.4 + 47.8	703.4 + 86.6	703.5 + 76.1	0.738	< 0.001	0.639
TP	3.72 + 0.89	6.13 + 1.13	6.87 + 0.80	3.44 + 0.92	5.97 + 1.02	6.64 + 0.70	0.597	0.001	0.758
LF	2.12 + 1.14	5.46 + 1.03	6.18 + 0.77	1.76 + 1.41	5.42 + 0.89	6.08 + 0.61	0.679	<0.001	0.535
HF	0.93 + 1.74	4.39 + 1.45	4.47 + 1.47	0.47 + 1.31	3.92 + 1.47	4.41 + 0.63	0.493	<0.001	0.841
LF/HF	1.19 + 0.97	1.07 + 0.56	1.71 + 0.78	1.30 + 0.65	1.50 + 0.66	1.67 + 0.37	0.160	0.231	0.747
LF%	48.58 + 16.92	69.11 + 9.85	79.22 + 12.13	51.73 + 16.78	75.42 + 8.83	80.98 + 5.06	0.130	0.031	0.865
HF%	9.02 + 7.94	20.86 + 8.98	13.82 + 9.88	6.74 + 4.72	15.06 + 6.98	13.54 + 4.80	0.125	0.001	0.702

HRV = heart rate variability; RR (ms) = mean of R-R intervals; TP [ln(ms2)] = Total power; LF [ln(ms2)] = low frequency power; HF [ln(ms2)] = high frequency; LF/HF (ln ratio) = the ratio of LF to HF; LF% = LF in normalized unit; HF% = HF in normalized unit.

were significantly lower than during the resting period. However, no significant interaction effect was observed for all of the HRV parameters between the cupping and the control trials during the 20-minute recovery period. These results indicated that, although cupping had a tendency to increase the parasympathetic activity at rest (Table 4), it did not significantly affect the recovery status of the autonomic nervous system after the exercise (Table 5).

4 Discussion

The purpose of the present study was to compare between dry cupping and control conditions for changes in RPE, MDF of triceps EMG, cardiorespiratory responses, and the recovery of HRV following high-intensity exercise that causes muscle fatigue. We found that cupping had a positive effect on subsequent exercise performance after high intensity exercise. The main effect was to reduce the fatigue level as reflected by the RPE and MDF of the triceps, but there was no significant effect on cardiorespiratory responses during exercise or on autonomic nervous system changes during recovery.

Muscle fatigue can be monitored using surface EMG during dynamic muscle contraction [16]. Fatigue causes the power spectrum of the EMG signal to move toward lower frequencies soon after the beginning of muscular activity and much earlier before a force or torque decrement [17]. The power spectral shift to lower frequencies (MDF) during fatigue and its causes and mechanisms are well documented [18]. In this study, the upper-extremity exercise program used to induce muscle fatigue was determined with reference to a previous study by Bernasconi et al. (2007) [19]. This program was utilized so as to clearly observe the effect of cupping on fatigue responses. In this study, we observed that the MDF of triceps signals shifted to lower frequencies after this exercise program. These results were consistent with the results of Bernasconi et al. (2007). However, we further found that cupping seemed to inhibit the shift of triceps MDF to lower frequencies, such that the participants reduced their ratings of perceived exertion.

A previous study has shown that short-term parasympathetic reactivation is impaired after repeated high-intensity exercise [20]. Otherwise, it has been proved that cupping therapy restores sympathovagal imbalances by stimulating the peripheral nervous system [7]. Therefore, the present study sought to verify whether cupping is helpful for the recovery of parasympathetic activity after high-intensity exercise, which would in turn affect the performance of subsequent exercise. However, the results of the present study showed that dry cupping only had the tendency to slow down the heartbeat and enhance the parasympathetic activity at rest, but did not affect the recovery of parasympathetic activity after high intensity exercise. The results further showed that the effect of cupping on autonomic nervous activity in this study was significantly smaller than that reported by Arslan et al. (2014). We speculate that this difference was due to the difference in the cupping style. Cupping therapy is performed in two forms: dry cupping and wet cupping. Our study utilized dry cupping. In contrast, Aslan et al. (2014) utilized a form of invasive wet cupping to create a small amount of wound bleeding, but this method is not suitable for sports competitions. In this study, the stimulation of cupping might not be sufficient to affect the autonomic nervous system, and so the autonomic nerve-controlled cardiopulmonary system also did not exhibit significant effects resulting from the cupping.

This study had several limitations: (1) This was a preliminary study, and the results must be interpreted with caution due to the small sample size. (2) This study investigated only the responses to cupping in moderately trained people. Because elite athletes have a distinct genetic make-up and training history, it is uncertain whether cupping effectively reduces their fatigue level as reflected by the EMG MDF. (3) There are many muscle groups involved in the arm cranking movement. However, this pilot study only observed the triceps. Future investigations would need to study more muscle groups, such as the biceps and deltoids. (4) This study evaluated the effects of cupping only on the performance of the upper limbs. Previous studies, however, have confirmed that there are many differences between upper- and lower-extremity exercises [21, 22]. Consequently, the effects of cupping on the exercise performance of the upper extremities cannot be assumed to generalize to the lower extremities, and further evidence is required regarding the effects of cupping on that part of the body.

5 Conclusions

Dry cupping treatment between two exercise bouts significantly decreased the levels of muscle fatigue as reflected in EMG and RPE values. Nevertheless, cupping did not cause significant differences in cardiorespiratory responses during exercise, and had no effect on the post-exercise recovery of HRV. Therefore, dry cupping therapy might have the positive effect of increasing exercise performance capacity during repetitive bouts of exercise.

References

1. Bompa, T.O.: Theory and Methodology of Training: the Key to Athletic Performance. Kendall Hunt Publishing Company, Dubuque (1994)
2. Barnett, A.: Using recovery modalities between training sessions in elite athletes. Sports Med. **36**(9), 781–796 (2006)
3. Mehta, P., Dhapte, V.: Cupping therapy: A prudent remedy for a plethora of medical ailments. J. Tradit. Complement Med. **5**(3), 127–134 (2015)
4. Lauche, R., et al.: Effectiveness of home-based cupping massage compared to progressive muscle relaxation in patients with chronic neck pain—a randomized controlled trial. PLoS ONE **8**(6), e65378 (2013)
5. Hanan, S., Eman, S.: Cupping therapy (al-hijama): it's impact on persistent non-specific lower back pain and client disability. Life Sci J **10**(4), 631–642 (2013)
6. Mccabe, P.J.: Complementary therapies in nursing and midwifery: from vision to practice (2001)
7. Arslan, M., et al.: Wet cupping therapy restores sympathovagal imbalances in cardiac rhythm. J. Altern. Complement. Med. **20**(4), 318–321 (2014)
8. Lewis, M.J., et al.: Rate of reduction of heart rate variability during exercise as an index of physical work capacity. Scandinavian J. Med. Sci. Sports **17**(6), 696–702 (2007)
9. Micklewright, D., et al.: Blood lactate removal using combined massage and active recovery. Biol. Sport **23**(4), 315 (2006)
10. Rozenfeld, E., Kalichman, L.: New is the well-forgotten old: the use of dry cupping in musculoskeletal medicine. J. Bodywork Mov. Ther. **20**(1), 173–178 (2016)
11. Leicht, A., Sealey, R., Sinclair, W.: The reliability of VO2peak determination in healthy females during an incremental arm ergometry test. Int. J. Sports Med. **30**(07), 509–515 (2009)
12. Forman, D.E., et al.: Cardiopulmonary exercise testing: relevant but underused. Postgrad. Med. **122**(6), 68–86 (2010)
13. Bressel, E., et al.: The effect of handgrip position on upper extremity neuromuscular responses to arm cranking exercise. J. Electromyogr. Kinesiol. **11**(4), 291–298 (2001)
14. Kuo, T.B., et al.: Effect of aging on gender differences in neural control of heart rate. Am. J. Physiol. Heart Circ. Physiol. **277**(6), H2233–H2239 (1999)
15. Heart rate variability standards of measurement, physiological interpretation, and clinical use. Task force of the European society of cardiology and the North American society of pacing and electrophysiology. Eur. Heart J. **17,** 354–381 (1996)
16. Ament, W., et al.: EMG median power frequency in an exhausting exercise. J. Electromyogr. Kinesiol. **3**(4), 214–220 (1993)
17. Merletti, R., Roy, S.: Myoelectric and mechanical manifestations of muscle fatigue in voluntary contractions. J. Orthop. Sports Phys. Ther. **24**(6), 342–353 (1996)
18. Dimitrova, N., Dimitrov, G.: Interpretation of EMG changes with fatigue: facts, pitfalls, and fallacies. Journal of Electromyography and Kinesiology **13**(1), 13–36 (2003)
19. Bernasconi, S.M., et al.: Changes in oxygen uptake, shoulder muscles activity, and propulsion cycle timing during strenuous wheelchair exercise. Spinal Cord **45**(7), 468–474 (2007)
20. Buchheit, M., Laursen, P.B., Ahmaidi, S.: Parasympathetic reactivation after repeated sprint exercise. Am. J. Physiol. Heart Circ. Physiol. **293**(1), H133–H141 (2007)
21. Cerny, F.J., Ucer, C.: Arm work interferes with normal ventilation. Appl. Ergonomics **35**(5), 411–415 (2004)
22. Chen, C-L., et al.: Immediate effects of smoking on cardiorespiratory responses during dynamic exercise: arm vs. leg ergometry. Front. Physiol. **6**, 376 (2015)

Risk of Injuries Caused by Fall of People Differing in Age, Sex, Health and Motor Experience

Roman Maciej Kalina(✉) and Dariusz Mosler

Faculty of Physical Education, Department of Combat Sports,
Gdansk University of Physical Education and Sport, Gdansk, Poland
kom.kalina@op.pl, dariusz.mosler@gmail.com

Abstract. Falls are one of main causes of unintentional injuries. Measuring risk of injury caused by a fall is principal basis in teaching safe fall techniques to reduce such risk. There were several studies which measures risk of injuries caused by fall by susceptibility test of body injuries during a fall (STBIDF). All available works which involves measuring participants by STBIDF were collected and screened. 527 participants in 18 different groups were tested. The lowest indicator were noticed in group of karate athletes on advanced level (SBIDF = 0.2), while the highest value were shown by group of people with intellectual disability (SBIDF = 11.12). Overall risk of injury level of tested people is high. Lower risk is connected to physical activity and experience in martial arts. Higher risk of injury is connected with low physical activity level and co-occurrence of different forms of disabilities.

Keywords: Disability · Injury risk · Martial arts · Safe fall

1 Introduction

Falls are one of main causes of unintentional injuries. Most studies focus on falling risk only, but accidental fall is unpredictable phenomena for every people. Fall risk assessments are often composed by different facilities and hospitals based on risk of fall factors specific to their patient's characteristic and environment [1]. Most of strategies focus on reducing risk of fall by providing secure environment by eliminating risk factors such as slippery floor [2], items that someone may tip over and such [3]. All renown physiotherapy strategies involves improving strength, balance or gait indicators [4]. All those programs are insufficient, and despite applying all those strategies, falls still occur, even in secure environment such as hospitals [5].

There is no possibility of completely eliminating falls from human life. Even young, healthy men may fall and get an injury or even loose his life. Not to mention people with different kinds of disabilities and diseases, which involves neurological [6] and musculoskeletal dysfunctions [7] that disturbs walking pattern and body control during a fall. Most of papers focus on classifying different groups of disability to groups of higher risk of injuries [8], but such studies are not specific, which body parts are the most vulnerable and how to improve motor control during a fall. Such enigmatic statement

T. Ahram (ed.), *Advances in Human Factors in Sports, Injury Prevention and Outdoor Recreation*, Advances in Intelligent Systems and Computing 603,
DOI 10.1007/978-3-319-60822-8_8

as “group of higher risk of injury” often inclines being more careful with such people. And those kind of treatment often restricts them from many activities, reducing quality of life, mostly by staying at home [9]. But still, they may fall and risk their health and life because of lack of knowledge, how to defend themselves from such phenomena. Starting from that paradigm, there is clear way to an assumption, that fall is inevitable in some way, and even if strategies that reduces such risk are important, it seems to be not enough. Fall may be an inevitable phenomena, but it does not mean, that every fall needs to end up with an injury [10]. Therefore, assessing injury risk of different body parts due to a fall seems to be justified.

The aim of this study is a review knowledge about risk of injuries for different groups of people.

2 Materials and Methods

Susceptibility test of body injuries during a fall (STBIDF) was used [11]. Test is composed of three tasks, which assess way of defending parts of the body, which are most vulnerable to injuries during a fall (legs, hands, head and hips). Each improper collision of those particular body parts with the ground during change of position from vertical to horizontal (laying backward) were documented on score sheet (Table 1) with corresponding value (“0” for lack of error, “1” for first degree error and “2” for second degree error). General score obtained by participant indicates susceptibility of body injury during a fall (SBIDF). People who gain 0 points have low SBIDF, 1–3 points indicates average, 4–8 high and 9–14 very high SBIDF. There were several studies which measures risk of injuries caused by fall by susceptibility test of body injuries during a fall (STBIDF). The aim of this study is to obtain a knowledge about risk of injuries for different groups of people.

Table 1. Score sheet for documentation of susceptibility of body injuries during a fall.

Body part	Task 1			Task 2 2			Task 3			Sum of points	SBIDF
legs	-			-			0	1	2		
hips	0	1		0	1		0	1			
hands	0	1	2	0	1	2	0	1	2		
head	0	1		0	1		0	1			
Sum of points										Overall SBIDF:	

All available works which involves measuring participants by STBIDF were collected and screened. 9 papers were selected. Specific groups were extracted and categorized. Due to lack of specific data, only overall STBIDF result was taken into account. Different body parts could not be analyzed, because it were not described in every selected paper. Beside of that, number of participants, specificity of each group, age, risk of an injury and physical activity were described. Papers were organized in order from the lowest susceptibility to the highest.

3 Results

Through 9 collected papers, 527 participants in 18 different groups were tested. Mean susceptibility of body injuries during a fall (SBIDF) indicator value was 5.54. The lowest indicator were noticed in group of karate athletes on advanced level (SBIDF = 0.2), while the highest value were shown by group of people with intellectual disability (SBIDF = 11.12). Mean age of participants were 21.3 years old. The lowest indicator with associated with physical activity and martial arts experience. SBIDF indicator were higher for group with different impairments, despite attending different physical activities. None of presented groups shows low risk of an injury level (Table 2).

Table 2. Summary of all extracted groups, organized by SBIDF indicator value.

Groups	N	Age(x)	SBIDF	Risk level	Physical activity
Karatekas on advanced level (3 kyu to dan) [12]	10	29.4	0.2	Average	Active: martial arts
Male physical education students [13]	68	25.4	0.65	Average	Active
Women judoists [14]	13	20.9	1.31	Average	Active: martial arts
Men judoists [14]	33	20.2	1.57	Average	Active: martial arts
Karatekas on intermediate level (4 to 7 kyu) [12]	16	14.1	2.38	Average	Active: martial arts
Untrained women [14]	35	21.8	4.37	High	Non active
Untrained men [14]	40	22.5	4.92	High	Non active
Young girls [15]	29	12.3	5.35	High	Occasionally
Young boys [15]	37	13.5	5.5	High	Occasionally
Karatekas (beginners) (8 to 9 kyu) [12]	16	12.9	5.56	High	Active: martial arts
Amputee football players [16]	7	26.4	6.85	High	Active
Young boys with visual impairment [15]	24	13.1	7.97	High	Active
Young girls with visual impairment [15]	27	13.6	7.98	High	Partially
Young boys [17]	53	10 to 12	8.28	High	School or extra activity
Female students [11]	68	21.26	8.36	High	Active
Young girls [17]	35	10 to 12	8.69	High	School or extra activity
People with mental illness [18]	8	48.25	8.75	High	Low
People with intellectual disability [19]	8	45.83	11.12	Very high	Low

4 Discussion

Inability to control body during a fall occurs not only among people with impairments, but also among adolescents and healthy, physically active adults. Without proper training, such inability will maintain through whole life, and will be increasing with an age and co-occurring illnesses. Therefore, teaching safe fall techniques seems to be crucial in reducing such risk [10]. Including safe fall techniques in normal exercise routine as part of health-related training [20, 21].

Traditional teaching recommendations for safe fall techniques (judo, jujitsu, aikido, hapkido etc.) are based on strict forms. Fun forms of martial arts and simulation situations of balance loss and fall [22] applied during teaching safe fall results in similar motor adaptation effects [23]. However, they increased attractiveness of training session and could be applied during occasional, recreation activity. STBIDF has a merit, that it can be included in those form of exercises. Capable designer of such exercises (with maintaining required STBIDF task order) can diagnose a person without stress and restrictions (especially mental) characteristic for an atmosphere of formal motor testing. This is significant remark from perspective of diagnosing susceptibility of body injuries during a fall for people with intellectual dysfunctions.

5 Conclusion

The overall risk of injury level of tested people is high. Lower risk is connected to physical activity and experience in martial arts. Higher risk of injury is connected with low physical activity level and co-occurrence of different forms of disabilities. While complying with recommended methodological criteria STBIDF can be widely used in the diagnosis of the susceptibility of body injuries during a fall of people who understand spoken words to them instruction (command).

References

1. Perell, K.L., Nelson, A., Goldman, R.L., Luther, S.L., Prieto-Lewis, N., Rubenstein, L.Z.: Fall risk assessment measures: an analytic review. J. Gerontol **56**, 761–766 (2001)
2. Hauer, K., Lamb, S.E., Jorstad, E.C., Todd, C., Becker, C.: Systematic review of definitions and methods of measuring falls in randomised controlled fall prevention trials. Age Ageing **35**, 5–10 (2006)
3. Tinetti, M.E.: Multifactorial fall-prevention strategies: time to retreat or advance. J. Am. Geriatr. Soc. **56**, 1563–1565 (2008)
4. Sherrington, C., Michaleff, Z.A., Fairhall, N., Paul, S.S., Tiedemann, A., Whitney, J., Cumming, R.G., Herbert, R.D., Close, J.C.T., Lord, S.R.: Exercise to prevent falls in older adults: an updated systematic review and meta-analysis. Br. J. Sports. Med., 1–10 (2016)
5. Chan, C.-H.: Gau, S.S-F., Chan, H.-Y., Tsai, Y.-J., Chiu, C.-C., Wang, S.-M., Huang, M.-L.: Risk factors for falling in psychiatric inpatients: a prospective, matched case-control study. J. Psychiatr. Res. **47**, 1088–1094 (2013)
6. Axer, H., Axer, M., Sauer, H., Witte, O.W., Hagemann, G.: Falls and gait disorders in geriatric neurology. Clin. Neurol. Neurosurg. **112**, 265–274 (2010)
7. Santello, M.: Review of motor control mechanisms underlying impact absorption from falls. Gait Posture **21**, 85–94 (2005)
8. Schwartz, A.V., Nevitt, M.C., Brown, B.W., Kelsey, J.L.: Increased falling as a risk factor for fracture among older women: the study of osteoporotic fractures. Am. J. Epidemiol. **161**, 180–185 (2005)

9. Bagalà, F., Becker, C., Cappello, A., Chiari, L., Aminian, K., Hausdorff, J.M., Zijlstra, W., Klenk, J.: Evaluation of accelerometer-based fall detection algorithms on real-world falls. PLoS ONE **7**, e37062 (2012)
10. Kalina, R.M., Barczyński, B.J., Jagiełło, W., Przeździecki, B., Kruszewski, A., Harasymowicz, J., Syska, J., Szamotulska, K.: Teaching of safe falling as most effective element of personal injury prevention in people regardless of gender, age and type of body build – the use of advanced information technologies to monitor the effects of education. Arch. Budo **4**, 82–90 (2008)
11. Kalina, R.M., Barczyński, B.J., Klukowski, K., Langfort, J., Gasienica-Walczak, B.: The method to evaluate the susceptibility to injuries during the fall – validation procedure of the specific motor test. Arch. Budo **7**, 203–216 (2011)
12. Adamczyk, J.G., Antoniak, B., Boguszewski, D., Siewierski, M.: The physical fitness and the safety falling skills of karatekas. J. Combat Sport Martial Arts **3**, 53–58 (2012)
13. Mroczkowski, A., Mosler, D.: Rotating training simulator as an assessment tool measuring susceptibility of the body injuries during the fall caused by an external force – validation procedure. In: Kalina, R.M. (ed.) Proceedings of the 1st World Congress on Health and Martial Arts in Interdisciplinary Approach, HMA 2015, 17–19 September 2015, Czestochowa, Warsaw, Poland (2015). Archives of Budo: 202
14. Boguszewski, D., Kerbaum, K.: Judo training as a means of reducing susceptibility to injury during a falls. Polish J. Sport. Med. **27**(3), 205–212 (2011)
15. Boguszewski, D., Zabłocka, M., Adamczyk, J.: Susceptibility to injury during a fall among blind children. Adv. Rehabil. **26**, 5–12 (2012)
16. Gąsienica-Walczak, B., Kalina, A.: Susceptibility of body injuries during a fall of people after amputation or with abnormalities of lower limb. In: Kalina, R.M. (ed.) Proceedings of the 1st World Congress on Health and Martial Arts in Interdisciplinary Approach, HMA 2015, 17–19 September 2015, Czestochowa, Warsaw, Poland (2015). Archives of Budo 193–195
17. Mroczkowski, A., Sikorski, M.M.: The susceptibility to body injuries during a fall and abilities related to motor coordination of children aged 10 to 12. Arch. Budo Sci. Martial Art Extreme Sport **11**, 65–71 (2015)
18. Mosler, D.: Changes of susceptibility of body injuries during a fall of patients with mental impairment participating for several months in special cognitive- behavioral therapy. In: Kalina, R.M. (ed.) Proceedings of the 1st World Congress on Health and Martial Arts in Interdisciplinary Approach, HMA 2015, 17–19 September 2015, Czestochowa, Warsaw, Poland (2015). Archives of Budo: 196–198
19. Mosler, D., Kmiecik-Małecka, E., Kalina, R.M.: Zmiany podatności na uszkodzenia ciała podczas upadku pacjentów z zaburzeniami psychicznymi objętych specjalną półroczną terapią poznawczo-behawioralną. Międzynarodowa Konferencja Naukowa Physiotherapy and Health Activity. Wychowania Fizycznego, Katowice (2014). In Polish
20. Kalina, R.M., Barczyński, B.J.: EKO-AGRO-FITNESS© original author's continuous program of health-oriented and ecological education in the family, among friends or individually implemented – the premises and assumptions. Arch Budo **6**, 178–184 (2010)
21. Kalina, R.M.: The profile of sense of positive health and survival abilities indices (subjective assessment) as a diagnostic tool used in health-related training. Arch. Budo **8**(3), 179–188 (2012)
22. Jagiełło, W., Kalina, R.M., Klimczak, J., Ananczenko, K.V., Ashkinazi, S., Kalina, A.: Fun forms of martial arts in positive enhancement of all dimensions of health and survival abilities. In: Kalina, R.M. (ed.) Proceedings of the 1st World Congress on Health and Martial Arts in Interdisciplinary Approach, HMA 2015, 17–19 September 2015, Czestochowa, Warsaw, Poland (2015). Archives of Budo: 32–39
23. Gąsienica-Walczak, B., Barczyński, B.J., Kalina, R.M., Kucio, C.: The effectiveness of two methods of teaching safe falls to physiotherapy students. Arch. Budo **6**(2), 63–71 (2010)

Physical Fitness and Exercise

Development of a Depth Camera-Based Instructional Tool for Resistive Exercise During Spaceflight

Linh Vu[1(✉)], Han Kim[2], Elizabeth Benson[3], William Amonette[4], Andrea Hanson[5], Jeevan Perera[5], and Sudhakar Rajulu[5]

[1] Geologics, Houston, TX, USA
linh.q.vu@nasa.gov
[2] Leidos, Houston, TX, USA
[3] MEI Technologies Inc., Houston, TX, USA
[4] University of Houston - Clear Lake, Houston, TX, USA
[5] NASA Johnson Space Center, Houston, TX, USA

Abstract. Resistive exercise is essential to maintaining proper musculoskeletal health during spaceflight. Therefore crewmembers receive instruction from strength, conditioning, and rehabilitation specialists on proper exercise technique to maximize exercise effectiveness and prevent injuries. However, long duration missions can make real-time exercise instruction and feedback problematic. A depth camera-based software tool was developed to provide exercise instruction and feedback during the deadlift exercise. The software tool uses a machine learning algorithm to identify 5 common deadlift technique mistakes. A subset containing 2 subjects with no deadlift training experience were coached on the deadlift exercise and separated into 2 groups: experimental group using the software tool or a control group without the tool. Motion-capture data were collected to evaluate the kinematic characteristics between the test and control group. It was found that the software tool assisted with increased torso, hip, and knee joint angle consistency and improved form during deadlifts.

Keywords: Biomechanics · Motion learning · Exercise performance assessment · Virtual coaching

1 Introduction

Bone density and muscle loss present serious health issues for astronauts. In reduced gravity environments, the human body is not subject to the same stresses as they would be on Earth. As a result, crewmembers may lose an average of 1% to 2% bone mass each month in microgravity [1]. Significant bone loss can lead to the potential for injury during or after missions. Exercise has been used as a vital countermeasure to maintain proper musculoskeletal health. Thus, crewmembers engage in regular physical exercise in microgravity. However, it is very difficult to replicate weight-bearing exercise in a near weightless environment. Specially designed equipment such as the advanced resistive exercise device (ARED) has been developed to allow crewmembers to simulate free-weight exercises. The

T. Ahram (ed.), *Advances in Human Factors in Sports, Injury Prevention and Outdoor Recreation*, Advances in Intelligent Systems and Computing 603,
DOI 10.1007/978-3-319-60822-8_9

introduction of resistive exercise equipment aboard the International Space Station has shown an increased rate of new bone formation and bone density conservation [2].

Similar to free-weight exercises on Earth, improper exercise techniques may result in musculoskeletal injuries and can pose a large risk to crewmembers. For example, excessive torque and irregular motion can place unnecessary stress and cause damage on joints. Having good exercise form ensures that joints are stable and the correct muscles are fully used during the exercise. In the last few decades, roughly one fourth of the weightlifting injuries resulting in emergency room visit were attributed to improper use of training equipment [3].

Before flight, crewmembers receive exercise instruction from astronaut strength, conditioning, and rehabilitation specialists. Once crewmembers have left Earth, they have limited access to exercise instruction and feedback from ground support. As missions extend beyond low-Earth orbit, exercise feedback will become more difficult due to longer communication delays. There is a need for a tool that can provide real-time instruction and correctional feedback for crewmembers in the absence of human coaching to prevent avoidable injuries and optimize overall muscle strength outcomes.

Virtual training environments can address these challenges by providing autonomous exercise guidance and feedback. In these systems, full-body motion is often evaluated by integrated camera or sensor-based motion tracking systems. Virtual personal trainers and coaches have been successfully used in physical therapy [4] and rehabilitation [5] applications. With strict volume limitations on current spacecraft, a crucial requirement is a standalone, self-contained, and small-form factor device with minimum setup time. Microsoft Kinect for Xbox One has been identified as a viable solution that can satisfy the requirement. The Microsoft Kinect uses a depth sensor with infrared projections to generate a depth map and detect movement. A random forest algorithm then determines body position and the 3D location of 26 skeletal joints during movement without requiring physical markers.

2 Objective

The purpose of this study is to develop and evaluate a virtual coaching tool capable of providing real-time instruction and feedback for the deadlift exercise. An instructional system was developed using a Microsoft Kinect depth-camera device that provides markerless 3-D whole-body motion capture at a small-form factor and minimal setup effort. We hypothesized that subjects using the newly developed instructional software tool would perform the deadlift exercise with more optimal kinematics and consistent technique than subjects without the instructional software.

3 Methods

3.1 Software Instruction Tool

A software tool was created to provide instruction and algorithmically evaluate deadlifts. The Kinect is placed approximately 6 feet in front of the user and 6 in above the ground. Joint-center position data were acquired at a sampling rate 30 frames per second. The tool provided a real-time sagittal projection of the limb and joint positions (Fig. 1). After every deadlift, the user is provided with verbal and visual instruction on how to improve the next repetition.

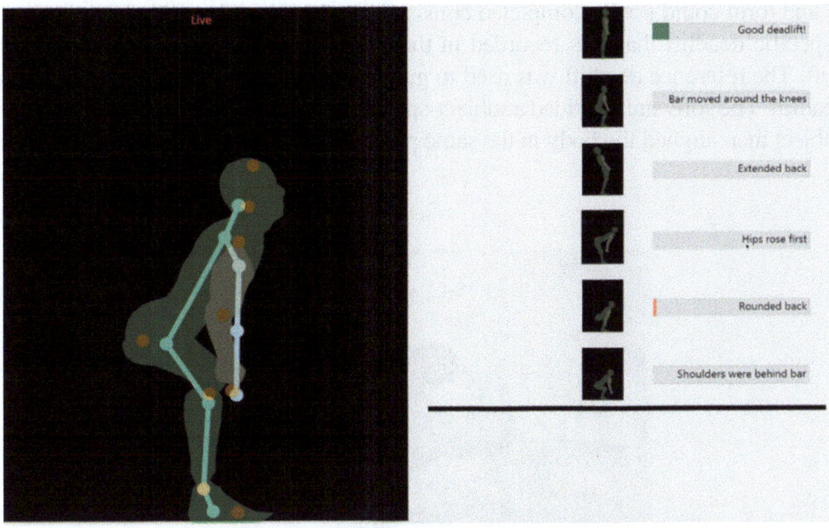

Fig. 1. Software tool interface

The software was pretrained with deadlift motions at various styles and speed, which was recorded with the Microsoft Kinect from a novice group of 20 participants. An Adaptive Boosting (AdaBoost) machine learning algorithm decomposed the recorded motions into feature vectors of body segment positions and velocities. AdaBoost is a sequential ensemble learning method that linearly combines multiple weak classifiers to form highly accurate classifiers to determine the proper classification [6]. An expert training coach visually classified the motions into either a "good deadlift" or 5 common error categories, which were statistically associated with the feature vectors. These common deadlift errors include rounding of the back, hips raising before the shoulders, lifting the weight bar around the knees, overextending the back at completion of the deadlift, and shoulders distal to the vertical plane of the weight bar. The trained software tool was able to classify newly observed motions and determine which categories the motion may fit best. Confidence level indices were reported for each category.

3.2 Procedure

A matched-pair randomized control study was conducted with subject pairs that were matched for gender and age. Twenty subjects that were recreationally active, but had no training experience with the deadlift, were invited to participate in the study. Written informed consent was obtained from each subject and all procedures were approved by the Johnson Space Center's Institutional Review Board (IRB). Two (both female) of the 20 subjects have participated in 4 total sessions that were separated by 1 or 2 weeks. Continued data collection is currently in progress.

Subjects were provided with standardized coaching instruction on the deadlift exercise in session 1. In session 2, a one repetition maximum determination (1-RM) was performed (Table 1). Weight was increased slowly on each subsequent lift until proper deadlift technique and form could not be completed consistently. Additionally, the subject performed a user-specific deadlift that was recorded in the software tool as the "model" or reference deadlift. The reference deadlift was used to guide users in the experimental group through the deadlift. The software provided a subject-specific silhouette in the basic start position and the subject then aligned the body in the same pose. A silhouette then repeated the reference

(a) (b) (c)

Fig. 2. Subject interacting with the software by: (a) aligning body joints in the proper position (b) following the motion of the silhouette (c) receiving verbal and visual feedback on motion

deadlift and the user was instructed to follow the motion (Fig. 2). Feedback is provided at the end of the deadlift.

Table 1. Subject demographics and one repetition maximum

Subject	Test group	Age	Weight (lb)	1-RM (lb)
C1	Control	26	166	132
E1	Experimental	26	148	145

An 8 camera, optical-passive Vicon motion-capture system (Vicon, Oxford, UK) was used to record the subject's motions in the last 2 sessions. Retro-reflective markers were adhered to the subject based on a modified Plug-in Gait marker set. Additional markers were placed on the iliac crest to assist with hip joint tracking. For session 3 and session 4, subjects were randomized and split equally into either an experimental group with the Kinect instructional tool or a control group without the tool. In session 3, subjects performed 5 sets of 3 deadlifts at 80% of the 1-RM. The same conditions and protocol were repeated in session 4.

3.3 Data Analysis

Biomechanical analysis was constrained to the concentric stage or lifting of the weight of the deadlift. In this current study, the kinematic characteristics of the experimental and control group were calculated and compared during the concentric stage of the deadlift. Specifically, the trajectories and angles of the lower extremity joints (torso, hip, and knee) were quantified to determine if using the software tool improves the efficacy and consistency of deadlift technique. Accuracy with respect to each subject's ideal or reference deadlift was assessed.

4 Results

The recorded joint angles at the torso, hip, and knee were averaged across deadlift trials and plotted on Fig. 3, in addition to the subject-specific "model" or reference deadlift. Motion was normalized to the corresponding duration. It is observed that a clear difference exists between the 2 subjects with respect to joint trajectory. The subject in the experimental group performed deadlifts with reasonably consistent form and the joint trajectories closely matched with those in the reference deadlift, particularly for the torso flexion/extension angle. Additionally, joint velocities are similar to the reference deadlift. In contrast, the deadlifts performed by the subject in the control group show a larger deviation from the model deadlift joint angles (torso flexion/extension angle). Also, the control group tends to show a greater variability in joint angle trajectory (knee flexion/extension).

Joint angle excursion ranges for subject E1 (experimental) show better agreement to the subject-specific reference deadlift than subject C1 (control). In the reference deadlift, Subject E1 have peak joint angles [max – min] ranges of [29.9–4.3], [95.7–4.3], and [99.5–12.1] degrees for torso, hip, and knee flexion/extension, respectively (Table 2). During the

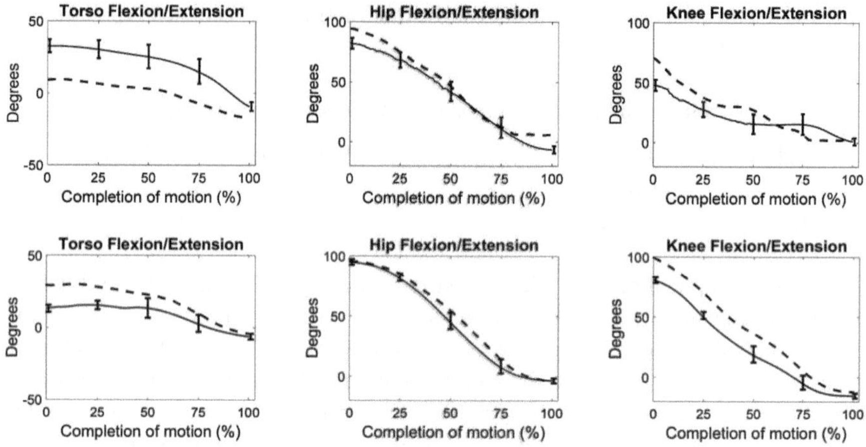

Fig. 3. Normalized time series of reference deadlift joint angles (dotted line), trial average (solid line) deadlift joint angles, and 1 standard deviation error bars for subject C1 (*top row*) and subject E1 (*bottom row*)

deadlift trial sessions for subject E1, the average excursion ranges are [17.9–6.1], [95.2–3.6], and [82.0–15.1] degrees, showing differences of 0.7° to 17° approximately.

Table 2. Average peak angle, excursion ranges, and standard deviation (parenthesis) for model deadlift and average of deadlift trial repetitions.

Joint angle	Subject	Type	Excursion range	Maximum joint angle	Minimum joint angle
Torso flexion/ extension	C1	Model	26.7	17.1	−9.6
		Trial	40.4 (6.3)	31.3 (6.3)	−9.1 (4.9)
	E1	Model	34.2	29.9	−4.3
		Trial	23.9 (3.9)	17.9 (5.7)	−6.1 (3.6)
Hip flexion/ extension	C1	Model	88.3	94.2	5.9
		Trial	86.8 (4.6)	81.2 (5.1)	−5.6 (4.1)
	E1	Model	100.0	95.7	−4.3
		Trial	98.8 (1.3)	95.2 (2.2)	−3.6 (1.8)
Knee flexion/ extension	C1	Model	73.2	70.4	−2.8
		Trial	44.6 (14.6)	44.9 (10.4)	0.4 (6.3)
	E1	Trial	111.5	99.5	−12.1
		Model	97.1 (4.6)	82.0 (5.7)	−15.1 (2.1)

Unit: degrees

For subject C1, the corresponding ranges are [17.1–9.6], [94.2–5.9], and [70.4–2.8] degrees during the reference deadlift. However, during the deadlift trial sessions, subject C1 have joint angle ranges of [31.3–9.1], [17.9–6.1], and [95.2–3.6] degrees. The differences are approximately 0.5° to 76°.

A closer look at the timing of motion shows that the joint angles for subject E1 are more consistent and closely matches the model deadlift at each phase of the motion in comparison to subject C1 (Table 3). This trend is particularly evident at the beginning of the deadlift. At the initiation of the deadlift (0% motion duration) for subject E1, the torso hip, and knee joint angle differs from the model deadlift by 15.0°, 0.5°, and 17.5°, respectively. For subject C1, the corresponding difference are 12.2°, 12.9°, and 54.7°. For all joint angles, subject E1 also exhibits less variability between deadlifts than subject C1 at all phases of the motion. However, both subjects have greater and noticeable variability in the middle of the deadlift (25%, 50%, and 75% of motion duration).

Table 3. Subject joint angles and standard deviation (parenthesis) through completion of the deadlift.

Joint angle	Subject	Type	Phase (% of motion duration)				
			0%	25%	50%	75%	100%
Torso flexion/ extension	C1	Model	17.1	7.1	−3.1	−6.6	−9.6
		Trial	29.3 (11.3)	27.1 (10.9)	22.7 (9.6)	13.3 (7.7)	−7.3 (8.1)
	E1	Model	29.1	28.1	22.7	9.6	−4.3
		Trial	14.1 (6.8)	16.2 (6.9)	13.9 (5.3)	2.8 (4.3)	−6.1 (3.6)
Hip flexion/ extension	C1	Model	94.1	75.1	46.7	13.1	6.2
		Trial	81.2 (5.1)	69.3 (5.5)	42.5 (7.0)	12.4 (7.6)	−5.4 (4.2)
	E1	Model	95.7	85.8	54.6	14.1	−4.3
		Trial	95.2 (2.2)	82.0 (3.0)	45.4 (7.0)	8.0 (5.5)	−3.6 (1.8)
Knee flexion/ extension	C1	Model	70.4	38.1	27.6	7.4	3.4
		Trial	44.8 (10.5)	27.9 (5.7)	17.1 (3.7)	14.3 (9.1)	1.0 (6.5)
	E1	Trial	99.5	71.2	36.4	5.7	−12.1
		Model	82.0 (5.7)	51.7 (8.1)	19.7 (6.2)	−4.4 (5.4)	−15.0 (2.0)

Unit: degrees

5 Discussion

The subject with the software tool performed the deadlift with more consistent and proper technique than the subject in the control group. The joint trajectories for subject E1 were comparable to the model deadlift, and joint angle was similar at all phases of the motion. The differences between the experimental and control groups were more pronounced during the initiation of the movement. It was observed that subject E1 would initiate the deadlift with consistent form because of repeated alignment with the silhouette provided by the software. In contrast, the control subject was provided with no reference, and would regularly initiate the deadlift in a much more upright starting position. The deadlift was performed with considerably less knee flexion, which showed little resemble to a proper deadlift. As the control subject did not receive guidance from the software tool through the rest of the deadlift motion, the control subject also displayed erratic movement patterns.

Through the initial testing, several limitations with the software tool were discovered. Despite having good alignment with silhouette, the knee flexion/extension angle differed from the model deadlift by roughly 14°. It was observed that the subject had a wider stance throughout the data collection sessions than during the recording of the reference deadlift. A wider stance may cause the sagittal joint projection of the software to become distorted, causing improper knee and torso positioning. Additionally, the feedback portion of the software was often incorrect for subject E1. Based on trainer feedback, the subject was performing the deadlifts with proper form, yet the software was classifying the deadlift as if the subject had rounded the back. A wider stance may have contributed to the misclassification. Therefore additional training of the software with various forms and deadlift techniques is needed to increase the accuracy of the software and reduce prediction errors. Another issue is that the subject had a difficult matching the speed of the silhouette through the deadlift. The silhouette will need to be adjusted to better compensate for human reaction lag in future work.

Currently data have been analyzed for 2 subjects, but data collection is ongoing. Therefore it is still early to decisively evaluate the efficacy of the software tool at this stage. Confounding factors exist that may have resulted from individual subject differences such as motor learning aptitude. Biomechanical analysis was also limited to the lower extremities, but upper body kinematics do play a large role in proper deadlift technique. For example, shoulder flexion and shoulder joint position may indicate improper weight bar movement, causing unnecessary torque to the back. These issues will be reduced once additional data are collected and further analysis is performed. As the eccentric motion of the deadlift also could lead to injury if performed poorly and future work should explore training the software to evaluate the full deadlift motion.

Despite the limited data, the findings provided evidence that the software tool had a positive impact. The subject had a favorable response to the software and the learning curve was very brief. These findings provided an early understanding of the effectiveness of the software tool and areas for future development.

6 Conclusion

Data collection remains ongoing and the study will conclude when a total 20 subjects have completed the procedure. However, an early analysis has identified positive outcomes related to the subject's form with the software tool. The initial testing has demonstrated that the software improved consistency between deadlift sets and greatly assisted subject during the initiation of the deadlift. The recent findings are promising and could provide an effective resistive exercise coaching tool for use during spaceflight. After continued development, implementation of the software tool may greatly improve the efficacy and safety of crewmembers performing exercises.

Acknowledgments. The authors would like to acknowledge Joseph Barrera, Kellie Keener, and Rikki Patrick for their hard work during data collection and analysis.

References

1. LeBlanc, A., Schneider, V., Shackelford, L., West, S., Oganov, V., Bakulin, A., Voronin, L.: Bone mineral and lean tissue loss after long duration space flight. J. Musculoskelet. Neuronal Interact. **1**, 157–160 (2000)

2. Smith, S.M., Heer, M.A., Shackelford, L.C., Sibonga, J.D., Ploutz-Snyder, L., Zwart, S.R.: Benefits for bone from resistance exercise and nutrition in long-duration spaceflight: Evidence from biochemistry and densitometry. J. Bone Miner. Res. **27**, 1896–1906 (2012). doi:10.1002/jbmr.1647
3. Jones, C.S., Christensen, C., Young, M.: Weight training injury trends: a 20-year survey. Phys. Sportsmed. **28**, 61–72 (2000)
4. Li, B., Maxwell, M., Leightley, D., Lindsay, A., Johnson, W., Ruck, A.: Development of exergame-based virtual trainer for physical therapy using kinect. Games Heal **29**, 230–232 (2014). doi:10.1089/g4h.2012.0019
5. Sveistrup, H., McComas, J., Thornton, M., Marshall, S., Finestone, H., McCormick, A., Babulic, K., Mayhew, A.: Experimental studies of virtual reality-delivered compared to conventional exercise programs for rehabilitation. Cyberpsychology Behav. **6**, 245–249 (2003). doi:10.1089/109493103322011524
6. Freund, Y., Schapire, R.: A decision-theoretic generalization of on-line learning and an application to boosting. Comput. Learn Theory **55**, 119–139 (1995). doi:10.1006/jcss.1997.1504

The Effect of Awareness of Physical Activity on the Characteristics of Motor Ability Among Five-Year-Old Children

Akari Kamimura[1(✉)], Yujiro Kawata[2,3], Shino Izutsu[2,4], and Masataka Hirosawa[1,2,3]

[1] Institute of Health and Sports Science and Medicine, Juntendo University, 1-1, Hiraga-gakuendai, Inzai, Chiba 270-1606, Japan
a-kamimura@juntendo.ac.jp, rir_u222@yahoo.co.jp

[2] School of Health and Sports Science, Juntendo University, 1-1, Hiraga-gakuendai, Inzai-shi, Chiba 270-1606, Japan
yuukawa@juntendo.ac.jp

[3] Graduate School of Health and Sports Science, Juntendo University, 1-1, Hiraga-gakuendai, Inzai-shi, Chiba 270-1606, Japan
hiro-ma@b-star.jp

[4] Faculty of Sports and Health Sciences, Japan Women's College of Physical Education, 8-19-1, Kitakarasuyama, Setagaya-Ku, Tokyo, 157-8565, Japan
izutsu.shino@jwcpe.ac.jp

Abstract. We examined the relationship between the types of awareness about physical activity and motor ability among Japanese 5-year-old children. Data were collected from 169 five-year-old children (85 boys and 84 girls). We found that children's awareness of their own physical activity (liking, enjoyment, and the confidence in physical activity as their physical competence) and actual motor ability (20-m sprint, standing broad jump, throwing a tennis ball) were weakly associated. Moreover, in classifying children's awareness of physical activity, we extracted four clusters of awareness characteristics. Notably, we found that some children with negative awareness did not have poor motor ability, whereas some children with positive awareness did not have good motor ability. Understanding these phenomena would be essential for building exciting environments that help all children view physical activity as play.

Keywords: Motor ability · Awareness of physical activity · Children

1 Introduction

In recent years, the decreases evident in children's motor ability, and changes in playmates, play places, and playtime have been regarded as important social issues [1], both in Japan and in many other developed countries [2–6]. Indeed, in other countries, these issues have been comprehensively considered in relation to obesity or chronic diseases in childhood [7, 8]. During early childhood, children learn fundamental motor skills that serve them across the lifetime through engaging in physical activity as play [9, 10].

T. Ahram (ed.), *Advances in Human Factors in Sports, Injury Prevention and Outdoor Recreation*, Advances in Intelligent Systems and Computing 603,
DOI 10.1007/978-3-319-60822-8_10

Furthermore, children's physical activity has important implications for their cognitive and social development [11]. Therefore, preventing problems with physical activity in early childhood would be important for ensuring people's lifetime health.

In Japan, the FY2015 Survey on Physical Fitness and Motor Abilities reported that children's motor ability has been declining since 1985 [12]. Additionally, the implementation status of physical activity has become increasingly polarized, with children either engaging in habitual exercise or having no such habit at all; children's motor ability has become similarly polarized, and, in fact, the number of children with low motor ability has been increasing. This increasing prevalence of low motor ability was regarded as not completely unrelated to the polarization [12]. In addition, Sugihara et al. found that a miniaturized exercise guide by a teacher or coach aimed at the practice of a specific physical activity actually impaired the motor ability of kindergarteners [13]. This suggests that coaching aimed at the practice of a specific physical activity might block children's development [14] and emphasizes the importance of inspiring enjoyment through various movements as play.

Given this background, several guidelines have been proposed in Japan to help arrest the decline in children's motor ability and improve their attitude towards physical activity [15]. In particular, in 2013, the "Physical Activity Guideline for Japanese Young Children" was developed to provide specific advice on the types of exercise that should be performed [16]. Specifically, it reported on the problems related to physical activity in children, the importance of physical activity in childhood, and specific methods for improving physical activity that can be used by parents or teachers. These guidelines particularly emphasized the importance of encouraging children to enjoy the physical activity that they do engage in.

Thus, despite the growing importance of promoting children's physical activity and enjoyment of said activity, current guidelines and existing research lack an understanding of the association between children's attitudes toward physical activity (i.e., their feelings about physical activity) and motor ability. Therefore, the purpose of this study was to examine the relationship between types of awareness about physical activity and motor ability among Japanese 5-year-old children.

2 Methods

2.1 Participants

We collected data from 169 five-year-old children (85 male and 84 female) from public and private kindergartens in Tokyo, Japan.

2.2 Measurements

We collected data on children's basic characteristics (age, and sex), children's motor ability (performance during on a 20-m sprint, standing broad jump, and throwing a tennis ball), and awareness of his/her own physical activity (liking, enjoyment, and the confidence in physical activity as their physical competence).

Basic Characteristics. We collected information on the children's sex, age, and academic grade from the kindergarten teachers using a questionnaire.

Motor Ability. For the 20-m sprint, we prepared a 25-m straight alley with goal lines at 20-m (true goal line) and 25 m (fake goal line) to help children successfully navigate the 20-m sprint course. The measurer stood at the 25-m goal line and recorded each child's time from the start cue ("set and go") to the moment the child crossed the 20-m line.

For the standing broad jump, we set a 3-m tape measure on the floor and drew a balk line. We then instructed the children to jump as far as they could from the scratch line, using both their right and left feet at the same time. We recorded each child's distance of jump.

Finally, for the tennis ball throwing, we instructed the children to throw the tennis ball as far as possible using their dominant hand and recorded each child's throw.

Before each test of motor ability, we demonstrated to the children how each test was performed. Participants performed each test twice, and the best score was used in the analysis. All measurements were performed at around 10:00 am.

Awareness of Physical Activity. Children's awareness of physical activity was measured through individual interviews using illustrations. Specifically, we assessed children's liking of physical activity, enjoyment of physical activity, and confidence as their physical competence in performing physical activity.

To measure liking and enjoyment of physical activity, we showed the children an A3 board on which we had drawn a number line ranging from 1 to 10, with a sad child's face placed beside 0 and a smiling child's face beside 10. We then asked the children to answer whether they agreed or disagreed with the following questions using the numbers on the line: "Do you like physical activities?" (for measuring liking of physical activity) and "Do you enjoy physical activities?" (for measuring enjoyment of physical activity) using the above-mentioned board.

To measure the confidence in physical activity as their physical competence, we showed children another whiteboard depicting five running children competing in a race, with each child labeled with a score from 1 to 5 (on a 5-point Likert scale). We then asked children to choose which number best represented how they would finish if they were competing in the race (i.e., 1 = worst and 5 = highest).

2.3 Data Analysis

We conducted a correlational analysis, a hierarchical cluster analysis, and one-way analysis of variance to determine the children's types of awareness of their own physical activity, and how these types related to their motor ability. Statistical significance was set at $p < .05$.

3 Results and Discussion

3.1 Correlations Between Awareness of Physical Activity and Motor Ability

To confirm the relationship between children's awareness of physical activity and motor ability, we calculated Pearson's product-moment correlation coefficients. The results are shown in Table 1. We found no significant correlation between liking physical activity and the 20-m sprint performance; however, liking did have weak negative correlations with tennis ball throwing ($r = -.251$, $p < .01$) and standing broad jump performance ($r = -.234$, $p < .01$). For enjoyment of physical activity, we observed no significant correlation with the 20-m sprint performance. However, there were again weak negative correlations between enjoyment and tennis ball throwing ($r = -.263$, $p < .01$) and standing broad jump performance ($r = -.246$, $p < .01$). Finally, we found weak negative correlations between confidence in physical activity and performance on the 20-m sprint ($r = -.261$, $p < .01$), as well as weak positive correlations with performance on tennis ball throwing ($r = .292$, $p < .001$) and standing broad jump ($r = .250$, $p < .01$).

Table 1. Matrix of correlations between awareness of physical activity and motor ability.

	Motor ability					
	20-m sprint		Tennis ball throwing		Standing broad jump	
Awareness of physical activity						
Liking	0.072	*n.s.*	−0.251	**	−0.234	**
Enjoyment	0.137	*n.s.*	−0.263	**	−0.246	**
Confidence	−0.261	**	0.292	***	0.250	**

n.s.: not significant, $^{**}p < .01$, $^{***}p < .001$

The weak negative correlations observed between liking or enjoyment and performance on tennis ball throwing and standing broad jump suggests that children who enjoy/like physical activities do not necessarily have good motor ability. Conversely, the weak negative correlation of confidence in physical activity with 20-m sprint performance and the weak positive correlations with tennis ball throwing and standing broad jump performance indicate that children who have high confidence in their physical activities tend to be good at accurately assessing their performance.

3.2 Classification of Children's Awareness of Physical Activity

To classify children's awareness of physical activity, we conducted a hierarchical cluster analysis (Ward's method) using the scores for children's awareness of physical activity. Ultimately, four clusters of awareness were extracted as follows: Cluster A was labeled "positive awareness" ($n = 118$, approx. 69.82% of the participants), Cluster B was labeled "loss of enjoyment" ($n = 19$, approx. 11.24% of the participants), Cluster C was labeled "dislike" ($n = 21$, approx. 12.43% of the participants), and Cluster D was labeled "negative awareness" ($n = 11$, approx. 6.51% of the participants) (Fig. 1).

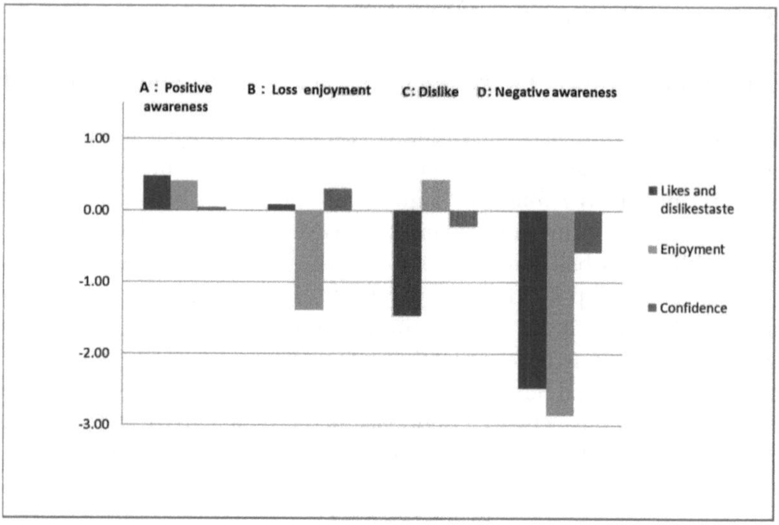

Fig. 1. Standardized scores of the four-clusters of awareness of physical activity among five-year-old children. (Note: Hierarchical cluster analysis (Ward's method) was conducted. The scores in the figure were standardized using the overall M and SD.)

We then conducted one-way ANOVAs to statistically clarify the differences between the four clusters (Clusters A, B, C, and D). Post hoc multiple comparisons were performed using Tukey's HSD test. A comparison of the awareness of physical activity scores among the four clusters is shown in Table 2.

Table 2. Comparison of the awareness of physical activity among the four-clusters.

Cluster	A: Positive awareness		B: Loss of Enjoyment		C: Dislike		D: Negative awareness		One-way ANOVA			Multiple comparison
	n = 118		n = 19		n = 21		n = 11					
	M	*SD*	*M*	*SD*	*M*	*SD*	*M*	*SD*	*F*	*P*	η^2	Tukey's HSD
Liking	9.71	.51	8.63	1.67	4.38	2.22	1.64	1.50	284.77	***	.70	D < C < B < A
Enjoyment	9.63	.50	5.16	1.74	9.67	.58	1.55	1.04	498.95	***	.81	D < B < A, C
Confidence	3.94	1.13	4.26	1.19	3.62	1.20	3.18	1.83	2.30	†	.00	D < B

†p < .10, ****p* < .001

For the liking score, the results showed that there were significant differences between Clusters A, B, C, and D ($F_{3, 165} = 284.77$, $p < 0.001$). Multiple comparisons indicated that Cluster D had significantly lower scores on liking than did the other three clusters (*ps* < 0.001). Furthermore, Cluster C demonstrated significantly lower liking scores than did Clusters A and B ($p < 0.001$), and Cluster B demonstrated significantly lower scores than did Cluster A ($p < 0.01$).

For enjoyment score, we again observed significant differences between the clusters ($F_{3, 165} = 498.96$, $p < 0.001$). Multiple comparisons indicated that Cluster D demonstrated significantly lower scores on enjoyment than did the other three clusters (*ps* < 0.001), while Cluster B demonstrated significantly lower scores than did Clusters

A and C ($p < 0.001$); however, there were no significant differences between Clusters A and C.

Finally, for confidence score, we observed significant differences between the clusters ($F_{3,\,165} = 2.30$, $p = .079$). The multiple comparisons indicated that only Cluster B demonstrated significantly higher scores on confidence than did Cluster D ($p < 0.10$); there were no significant difference among the other clusters.

Taken together, these results suggested that children's awareness of physical activity could be classified into four distinct types (positive awareness, loss of enjoyment, dislike, and negative awareness). Notably, the majority of children (around 70%), had positive awareness of physical activities. The remaining 30% of the children had negative awareness, loss of enjoyment, or dislike of physical activities. Therefore, in order to ensure lifetime mental and physical health for all children, it is important to consider how to improve this negative awareness.

3.3 Relations Between the Types of Children's Awareness and Motor Ability

We conducted a one-way ANOVA to determine the differences in children's motor ability according to the awareness type. Post hoc multiple comparisons were performed using Tukey's HSD test. The results of this comparison are shown in Table 3.

Table 3. Comparison of motor ability among the four-clusters.

Cluster	A: Positive awareness		B: Loss of Enjoyment		C: Dislike		D: Negative awareness		One-way ANOVA			Multiple comparison
	n = 118		n = 19		n = 21		n = 11					
	M	*SD*	*M*	*SD*	*M*	*SD*	*M*	*SD*	*F*	*P*	η^2	Tukey's HSD
20 m sprint	5.32	0.44	5.21	0.48	5.45	0.44	5.07	0.55	2.05	*n.s.*	.00	——
Tennis ball throwing	5.45	2.51	7.21	3.60	6.10	2.61	8.27	3.10	5.47	**	.01	A < B, D
Standing broad jump	98.59	15.41	103.42	15.99	99.43	14.07	113.14	21.14	3.17	*	.00	A < D

n.s.: not significant, *p < .05, **p < .01+

For the 20-m sprint, we observed no significant differences between the clusters ($F_{3,\,165} = 2.04$, $p > 0.1$), and the effect size was almost non-existent ($\eta^2 = .00$).

For tennis ball throwing, we did observe significant differences among the clusters ($F_{3,165} = 5.46, p < 0.01$), although the effect size was rather small ($\eta^2 = .01$). The multiple comparisons indicated that Cluster A showed significantly lower performance on the throwing test than did Clusters B ($p < 0.05$) and D ($p < 0.01$), while there were no significant differences among the other clusters.

Finally, for the standing broad jump, we observed significant differences among the clusters ($F_{3,\,165} = 3.17, p < 0.05$), although the effect size was again rather small ($\eta^2 = .01$). Multiple comparisons indicated that Cluster A demonstrated significantly lower scores on the standing broad jump than did Cluster D ($p < 0.05$); there were no other significant differences among the clusters.

These results suggest that the tennis ball throwing performance was higher among children who demonstrated the loss of enjoyment or negative awareness types of awareness, in comparison to those with the positive awareness type. Similarly, the standing broad jump performance among children with negative awareness was higher than that among children with positive awareness. However, considering the sample size as well as the effect size, these findings cannot be considered conclusive. The reason for the higher physical activity performance in children with negative awareness of loss of enjoyment may be that they do not derive positive feelings from play-based physical activity with same-age peers, having already obtained higher performance in physical activity through extracurricular sports lessons or playing with older siblings.

In Japan, the participation rates in sports lessons among preschool children are 18.4% at 3 years of age, 29.1% at 4 years of age, and 58.5% at 5 years of age [17]; indeed, the participation rate in sports lessons is higher for 5-year-olds than the participation rates in arts lessons or cram school. This participation rate in sports lessons is increasing annually [18]. Participation in sports lessons would give children the opportunity to actively move their body. However, they do not necessarily lead to greater spontaneous engagement in physical activity, which is important for younger children [14]. Therefore, we must further explore the reasons for the above results and help resolve the problems related to them to build an exciting environment for play-based physical activity for all children.

4 Conclusions

In conclusion, the awareness of physical activity was only weakly associated with motor ability as measured by 20-m sprint, tennis ball throwing, and standing broad jump performance. Moreover, we found that some children with negative awareness did not have poor motor ability, while, conversely, some children with positive awareness did not have good motor ability.

References

1. Nakamura, K.: Transfiguration of children's play. J. Health Phys. Educ. Recreation **49**, 25–27 (1999). (in Japanese)
2. Montgomery, C., Reilly, J.J., Jackson, D.M., Kelly, L.A., Slater, C., Paton, J.Y., Grant, S.: Relation between physical activity and energy expenditure in a representative sample of young children. Am. J. Clin. Nutr. **80**, 591–596 (2004)
3. Reilly, J.J., Jackson, D.M., Montgomery, C., Kelly, L.A., Slater, C., Grant, S., Paton, J.Y.: Total energy expenditure and physical activity in young Scottish children: mixed longitudinal study. Lancet **363**, 211–212 (2004)
4. Finn, K., Johannsen, N., Specker, B.: Factors associated with physical activity in preschool children. J. Pediatr. **140**, 81–85 (2002)
5. Pate, R.R., Pfeiffer, K.A., Trost, S.G., Ziegler, P., Dowda, M.: Physical activity among children attending preschools. Pediatrics **114**, 1258–1263 (2004)
6. Williams, H.G., Pfeiffer, K.A., O'neill, J.R., Dowda, M., McIver, K.L., Brown, W.H., Pate, R.R.: Motor skill performance and physical activity in preschool children. Obesity **16**(6), 1421–1426 (2008)

7. Ogden, C.L., Carroll, M.D., Curtin, L.R., McDowell, M.A., Tabak, C.J., Flegal, K.M.: Prevalence of overweight and obesity in the United States, 1999–2004. JAMA **295**, 1549–1555 (2006)
8. Ogden, C.L., Troiano, R.P., Briefel, R.R., Kuczmarski, R.J., Flegal, K.M., Johnson, C.L.: Prevalence of overweight among preschool children in the United States, 1971 through 1994. Pediatrics **99**, E1 (1997)
9. Gabbard, C.P.: Lifelong Motor Development, 5th edn. Pearson, San Francisco (2008)
10. Piek, J.P., Hands, B., Licari, M.K.: Assessment of motor functioning in the preschool period. Neuropsychol. Rev. **22**, 402–413 (2012)
11. Diamond, A.: Interrelated and interdependent. Dev. Sci. **10**, 152–158 (2007)
12. Ministry of Education, Culture, Sports, Science and Technology: Results of the FY2015 survey on physical fitness and motor abilities. http://www.mext.go.jp/en/news/topics/detail/__icsFiles/afieldfile/2015/07/02/1359170_3.pdf
13. Sugihara, T., Yoshida, I., Mori, S.: Relationship between motor ability, exercise instruction, and personality among Japanese young children. J. Health Phys. Educ. Recreation **60**(5), 341–347 (2010). (in Japanese)
14. Kondo, M.: The Growing of Exercise and Mind Among Young Children. Sekaibunka Publishing, Tokyo (1994). (in Japanese)
15. Science Council of Japan: The basic guidelines for good practice of exercise and sports activities in order to make an energetic kid (2011). (in Japanese). http://www.scj.go.jp/ja/info/kohyo/pdf/kohyo-21-t130-5-1.pdf
16. Ministry of Education, Culture, Sports, Science and Technology: Physical activity guideline for Japanese young children (2013). http://www.mext.go.jp/a_menu/sports/undousisin/1319772.htm
17. Benesse Educational Research and Development Institute: The 1st the investigation related to educational activity outside a school (2009). http://berd.benesse.jp/shotouchutou/research/detail1.php?id=3264
18. Ministry of Education, Culture, Sports, Science and Technology: Report of investigation of actual situation concerning related to learning activity outside a school among children. http://www.mext.go.jp/b_menu/houdou/20/08/08080710.htm

Effect of Relative Age on Physical Size and Motor Ability Among Japanese Elementary Schoolchildren

Yujiro Kawata[1,2](✉), Akari Kamimura[3], Shino Izutsu[4], and Masataka Hirosawa[1,2,3]

[1] Department of Health and Sports Science, Juntendo University, 1-1, Hiraga-gakuendai, Inzai-shi, Chiba 270-1695, Japan
yuukawa@juntendo.ac.jp

[2] Graduate School of Health and Sports Science, Juntendo University, 1-1, Hiraga-gakuendai, Inzai-shi, Chiba 270-1695, Japan

[3] Institute of Health and Sports Science and Medicine, Juntendo University, 1-1, Hiraga-gakuendai, Inzai-shi, Chiba 270-1695, Japan

[4] Japan Women's College of Physical Education, 8-19-1, Kitakawasuyama, Setagaya-ku, Tokyo 157-8565, Japan

Abstract. A child born soon after the designated cut-off date ("early-born") may benefit by up to a full year in physical and psychological development, unlike one born just before the cut-off date ("late-born"). This phenomenon—called the "relative age effect"—causes inequalities in education and impacts the psychological aspects of children's sports participation. However, its influence on physical aspects has not been clarified. Therefore, we examined its impact on physical size and motor ability among Japanese elementary schoolchildren (384 males and 360 females). We collected demographic data and measured physical size and motor abilities. Analysis of covariance showed that early-born children scored significantly higher than late-born children across variables, indicating the effect of relative age on physical aspects of sports participation and the possible advantage of early-born over late-born children in sports achievement. We propose the need for acknowledging this phenomenon in educational settings among teachers and coaches.

Keywords: Relative age effect · Physical size · Motor ability · Elementary schoolchildren

1 Introduction

The "Matthew effect"—when social advantage leads to further advantages or disadvantages leads to further disadvantages through time—creates widening gaps between those who have more and those who have less [1]. This phenomenon is emerging in various fields such as science, technology, economy, politics, and education [2]. One of the factors causing this phenomenon is known as "relative age."

Although school year systems were initially employed to avoid large age differences among students, a child born soon after a designated cut-off date (an "early-born"

T. Ahram (ed.), *Advances in Human Factors in Sports, Injury Prevention and Outdoor Recreation*, Advances in Intelligent Systems and Computing 603,
DOI 10.1007/978-3-319-60822-8_11

child) may benefit by up to a full year in his/her physical and psychological development when compared to one born just before the cut-off date (a "late-born" child). This phenomenon is called the "relative age effect" (RAE). It is defined as the consequence of age differences between individuals within the same grade level, either on sports teams or in school [3]. It causes inequalities in school settings.

The RAE on sports was first reported by Grondin et al. [4], according to the relative age in education. They addressed a possible relationship between relative age and sport participation in terms of the relationship between relative age and scholastic achievement. They argued that early-born children in the competition year possess a competitive advantage over their younger peers. Then, they demonstrated that the number of players born in the first months of the year was greater than that of players born in the last months of the year in the National Hockey League (NHL). They suggested that early-born children possess a competitive advantage over their younger peers. Following their report, a similar phenomenon was found in other sports such as soccer [5, 6] major league baseball [7], and basketball [8]. RAE was also observed in sports in Japan such as baseball, soccer, volleyball, Ekiden (track and field long-distance relay), basketball, and sumo wrestling, but was not observed in handball, golf, rugby, American football, or badminton [9]. In fact, the inverse situation was observed. The number of players born soon after the cut-off date is significantly smaller than that of players born just before the cut-off date in dance, gymnastics [10], and horse racing [9].

Findings indicate that, regarding the impact of RAE on education, children born within the initial months of an academic year excel over their younger counterparts. The difference of almost one year is also related to significant differences in children's cognitive development [11, 12]. Late-born children have more academic problems compared to their early-born classmates [13–15], as well as a lower attendance rates in school [16]. Late-born children are more likely to be classified as learning disabled [17, 18], and their academic achievement is significantly lower than that of their older classmates [19]. Therefore, late-born children are likely to have an educational disadvantage, at least in a young age. Of course, although, the difference in abilities may decrease and disappear when children grow up, the number of successful experiences may be different between early-born and late-born children. Appropriate support is necessary for children to achieve the maximum benefit from schooling, and a lack of knowledge regarding RAE in education is a serious problem.

To date, Kawata and his colleagues have investigated the effect of RAE on children's physical and psychological development. They [20, 21] found the effect on physical development (physical size and motor abilities) among four- and five-year-old Japanese kindergarten children and the effect on their teachers' evaluations. However, they [22] also found that RAE did not influence on the psychological development among Japanese kindergarten children of the same age, suggesting that children experience high enjoyment and confidence from participating in physical activities regardless of birth month. Kamimura et al. [23] indicated that the birth date could influence a teacher's evaluation, mediating in terms of physical size and motor abilities. Among children aged 6–12 years, Kawata et al. [24] discovered that RAE is evident in the psychological aspects of sports participation (enjoyment of physical activity and competence) in children. Thus, the effect on psychological aspects appears in children of elementary school age, possibly because of the development of children's ability to compare themselves with others or because they are receiving evaluations during elementary school.

Furthermore, RAE observed in psychological aspects may stem from RAE related to physical aspects. However, the impact of RAE on physical aspects (physical size and motor ability) among Japanese children aged 6–12 year has not yet been clarified. If the influence of RAE on physical aspects continues beyond elementary school, inequalities may appear in educational settings and situations.

Therefore, we examined the impact of RAE on physical size and motor ability among Japanese elementary schoolchildren. Taken together, identification of the existence and magnitude of RAE and the examination of potential consequences would be important for improving school education systems and policies.

2 Method

2.1 Participants

We collected data from 744 Japanese elementary school children (384 male, 360 female) from public elementary schools in Japan. Mean age was 8.76 ± 1.86 and age range was 6–12.

2.2 Measurements

We distributed a questionnaire to collect demographic information (gender, grade, and birth year/month/day), measured physical size (height, weight, and body mass index) and motor abilities (50 m sprint, standing broad jump, softball throw, grip strength, forward bend, quick jumps, sit-ups, and multistage fitness test).

Physical Size. Regarding physical size, we used a stadiometer to measure height and a weight scale to measure weight. Body mass index was calculated using the parameters of height and weight.

Motor Abilities

Speed. We used a 50 m sprint to assess speed. We created a 50 m straight alley and a goal line at 50 m. The measurer stood at the goal line and recorded children's time from the start cue ("set and go") to the moment they crossed the goal line.

Jumping Power. To evaluate jumping power, we employed the standing broad jump. We set a balk line in front of the outside sandbox. We instructed the children to jump into the sandbox as far as they could from the balk line, using both their right and left feet at the same time; then, we recorded their distance.

Throwing Power. To assess throwing power, we adopted an official softball (No. 1 size for 1st and 2nd grade students and No. 2 size for 3rd to 6th grade students) considering children's hand sizes for the softball throw. We instructed children to throw the ball as far as possible using the dominant hand; accordingly, we recorded the distance.

Grip Strength. To assess grip strength, we used a grip dynamometer. Our instructions were to squeeze it as much as possible with each hand. We recorded a grip force in each hand, and then we calculated the average of each hand's score.

Flexibility. To assess flexibility, we adopted the forward bend, instructing children to sit down on the floor barefoot and place the measurement equipment above their legs, bending forward at the waist. We recorded the distance they could bend forward.

Agility. To assess agility, we adopted the quick jump. We drew three parallel lines on the floor (a right line, center line, and left line to the side of children) and instructed the children to jump from the center line to the right line and return to the center line. Then, they were to jump from the center line to the left line and return to the center line. They were to repeat this exercise as rapidly as they could. We recorded the number of jumps within 20 s.

Muscle Endurance. To assess muscle endurance, we adopted sit-ups, instructing children to lay on the floor and sit up as many times as they could in 30 s. We recorded the number of sit-ups in the period.

Whole Body Endurance. To assess whole body endurance, we conducted multistage fitness test. We drew two parallel lines 20 m apart on the floor. We instructed children to go between the two lines as many times as they could, keeping pace with the tempo of the accompanying audio. We recorded the number of going back between the two lines within the designed time limit.

We showed some examples of how to perform all of these activities before the measurement trials. All measurements were taken midmorning.

2.3 Ethical Consideration

This study was approved by the Research Ethics Committee of the School of Health and Sports Science, Juntendo University. Prior to the study, we obtained permission from the school principals and board of education. Informed consent was obtained from parents of the participants. Each participant was made aware of his/her right to decline to cooperate at any time, even after consenting to participate, without repercussions.

2.4 Statistical Analysis

We divided participants into the following four groups based on their birth month: Group A (April 2 through June), Group B (July through September), Group C (October through December), and Group D (January through April 1). It should be noted that the Japanese school year runs from April 1 to March 31. Children born on April 1 were placed in Group D because that date is the cut-off date for grade placement according to Japanese law. Children in Groups A and B were considered early-born, and those in Groups C and D were considered late-born. First, we calculated the demographic information of participants: number of participants by grade level and gender and number of participants by groups based on birth date. Next, an analysis of covariance

(ANCOVA) was conducted to assess the differences in all variables among the four groups, with gender regarded as a covariate factor. We calculated η^2 as an effect size for each analysis. The effect size is regarded as a statistical measure of the strength of a phenomenon. Thus, we used the effect size to compare the strength of RAE in this study. The criteria of the effect size of η^2 were as follows: small effect ($\eta^2 = 0.01$), medium effect ($\eta^2 = 0.06$), and large effect ($\eta^2 = 0.14$) [25]. The statistical significance was set at $p < .05$.

3 Results and Discussion

3.1 Demographic Information of Participants

Table 1 shows the numbers and percentages of the participants by grade level and gender. The number and percentages of the participants were well balanced between grade levels.

Table 1. Number of participants by grade level and gender

	Gender				Total
	Male		Female		
	N	%	N	%	
1st grade	74	56.1	58	43.9	132
2nd grade	70	53.8	60	46.2	130
3rd grade	60	51.7	56	48.3	116
4th grade	70	50.7	68	49.3	138
5th grade	50	53.2	44	46.8	94
6th grade	60	44.8	74	55.2	134
Total	384	51.6	360	48.4	744

Table 2 shows the numbers and percentages of participants by birth date and gender. The number and percentages of participants were balanced between the four groups. However, there were significant differences in Groups C and D (i.e., fewer boys in Group C and fewer girls in Group D).

Table 2. Number of participants by groups based on birth date

	Gender				Total
	Male		Female		
	N	%	N	%	
Group A (April 2–June)	96	53.9	82	46.1	178
Group B (July–September)	86	51.8	80	48.2	166
Group C (October–December)	88	43.1	116	56.9	204
Group D (January–April 1)	114	58.2	82	41.8	196
Total	384	51.6	360	48.4	744

3.2 RAE and Physical Size

Table 3 shows the differences in height. In 1st grade, Groups A and B scored significantly higher than Group D. Groups A and C scored significantly higher than Group D in 2nd grade. Group A scored significantly higher than Groups C and D, and Group B scored significantly higher than Groups C and D in 3rd grade. Groups A and B scored significantly higher than Group D in 4th grade. Group A scored significantly higher than Groups B, C, and D in 5th grade. Group A scored significantly higher than Group D in 6th grade. The findings showed that early-born children scored higher than late-born children in all grades, indicating the existence of RAE for physical size (height).

Table 3. The relative age effect (RAE) on height.

	Group A		Group B		Group C		Group D		Main effect	Multiple comparison	Effect size
	M	SD	M	SD	M	SD	M	SD			
1st grade	118.8	2.7	117.2	6.0	116.1	4.4	113.5	4.0	**	A, B > D	0.11
2nd grade	125.0	3.9	122.4	6.0	123.3	4.3	119.5	4.9	***	A, C > D	0.15
3rd grade	131.8	4.7	131.0	5.0	126.4	3.5	124.4	4.0	***	A > C & D, B > C & D	0.30
4th grade	135.0	4.9	135.4	8.1	132.8	5.0	131.3	6.0	**	A & B > D	0.09
5th grade	140.3	4.7	138.6	4.4	138.8	7.2	138.9	6.1	*	A > B, C & D	0.03
6th grade	147.8	6.5	144.6	5.4	146.9	5.7	143.7	5.1	*	A > D	0.09

Note: Unit is centimeter. M = mean, SD = standard deviation.
$^{*}p < .05$, $^{**}p < .01$, $^{***}p < .001$

Table 4 shows the differences in weight. In 1st grade, no significant difference was found. Group C scored significantly higher than Group D in 2nd grade. Group A scored significantly higher than Groups C and D, and Group B scored significantly higher than Groups C and D in 3rd grade. No significant difference was found among children in 4th grade. Group A scored significantly higher than Group B in 5th grade, and Groups A and C scored significantly higher than Group D in 6th grade. The findings showed that early-born children scored higher than late-born children in some grades: therefore, RAE exists for physical size in terms of weight.

Table 4. The relative age effect (RAE) on weight

	Group A		Group B		Group C		Group D		Main effect	Multiple comparison	Effect size
	M	SD	M	SD	M	SD	M	SD			
1st grade	21.7	2.6	20.7	2.4	20.8	3.1	19.5	3.4			0.05
2nd grade	23.7	3.3	22.9	2.3	24.3	4.1	21.6	3.5	**	C > D	0.09
3rd grade	28.1	4.4	27.0	3.9	24.5	2.9	24.6	3.1	**	A > C & D, B > C & D	0.10
4th grade	31.4	6.7	31.8	9.9	28.1	4.0	28.3	5.4			0.06
5th grade	34.0	6.1	28.7	2.9	30.1	3.9	33.3	8.2	*	A > B	0.10
6th grade	38.5	5.4	37.0	6.7	38.9	7.6	34.3	4.7	**	A & C > D	0.09

Note: Unit is kilogram. M = mean, SD = standard deviation.
$^{*}p < .05$, $^{**}p < .01$

3.3 RAE and Motor Ability

Table 5 shows the difference in the 50 m sprint. In the 1st grade, Group B scored significantly lower than Group C. Group A scored significantly lower than Group D in the 4th grade. In other grades, no significant difference was found. This indicates that early-born children scored higher than late-born children did in some grades; therefore, the RAE exists for speed measured by 50 m sprints.

Table 5. The relative age effect (RAE) on 50 m sprint

	Group A		Group B		Group C		Group D		Main effect	Multiple comparison	Effect size
	M	SD	M	SD	M	SD	M	SD			
1st grade	11.3	0.9	11.1	0.6	11.9	1.5	11.4	0.9	**	B < C	0.09
2nd grade	10.1	1.0	10.3	0.6	10.3	1.0	10.8	0.9			0.06
3rd grade	10.5	1.2	10.0	0.6	10.5	0.8	10.4	0.7			0.09
4th grade	9.2	0.8	9.6	0.7	9.4	0.8	9.7	1.0	**	A < D	0.08
5th grade	9.1	1.0	8.8	0.5	8.8	0.5	9.0	0.9			0.03
6th grade	9.2	1.6	9.3	0.9	9.3	0.7	9.4	0.8			0.02

Note: Unit is seconds. M = mean, SD = standard deviation.
*$p < .05$, **$p < .01$

Table 6 shows differences among children for the standing broad jump. In the 1st year grade, Group A scored significantly higher than Group C. In other grades, no significant difference was found. Since early-born children scored higher than late-born children, the RAE appears to influence jumping power.

Table 6. The relative age effect (RAE) on standing broad jump

	Group A		Group B		Group C		Group D		Main effect	Multiple comparison	Effect size
	M	SD	M	SD	M	SD	M	SD			
1st grade	114.0	19.0	108.6	12.3	102.8	14.3	106.4	12.6	**	A > C	0.08
2nd grade	118.5	16.1	116.4	15.4	109.1	17.6	109.6	14.4			0.07
3rd grade	126.0	17.7	126.0	14.8	131.0	22.1	126.6	17.3			0.03
4th grade	132.0	12.1	126.2	14.7	126.2	18.8	124.0	17.7			0.04
5th grade	143.7	16.8	141.0	9.8	148.0	17.0	146.0	20.1			0.02
6th grade	141.1	23.9	145.1	20.1	143.3	17.3	139.9	23.8			0.01

Note: Unit is centimeter. M = mean, SD = standard deviation.
*$p < .05$, **$p < .01$

Table 7 shows the differences in softball throw. In 2nd grade, Group A scored significantly higher than Group B. Group A scored significantly higher than Groups C and D in 3rd grade. Group A scored significantly higher than Group C in 5th and 6th grades. In the other grades, no significant difference was found. The results showed that early-born children scored higher than late-born children in some grades, indicating that RAE influences throwing power.

Table 7. The relative age effect (RAE) on softball throwing

	Group A		Group B		Group C		Group D		Main effect	Multiple comparison	Effect size
	M	SD	M	SD	M	SD	M	SD			
1st grade	7.3	1.8	6.1	2.6	9.7	2.6	6.7	3.1			0.03
2nd grade	11.0	4.0	7.9	2.9	9.4	4.2	9.0	3.5	***	A > B	0.15
3rd grade	11.5	4.9	11.2	4.9	9.5	3.0	9.3	5.0	***	A > C & D	0.36
4th grade	15.8	7.0	14.8	6.9	14.1	6.7	15.0	6.5			0.03
5th grade	18.2	6.4	16.2	8.2	14.1	5.2	19.6	4.1	*	A > C	0.19
6th grade	22.2	9.9	18.5	7.7	17.1	7.7	19.8	9.9	*	A > C	0.06

Note: Unit is meter. M = mean, SD = standard deviation.
$*p < .05$, $***p < .001$

Table 8 shows the difference in grip strength. In 1st grade, Group B scored significantly higher than Group D. Groups A and B scored significantly higher than Group D in 2nd grade. Group A scored significantly higher than Group D in 4th and 6th grades. In the other grades, no significant differences were found. Findings showed that early-born children scored higher than late-born children in some grades, suggesting that RAE influences grip strength.

Table 8. The relative age effect (RAE) on grip strength

	Group A		Group B		Group C		Group D		Main effect	Multiple comparison	Effect size
	M	SD	M	SD	M	SD	M	SD			
1st grade	8.8	1.7	8.0	1.7	7.3	1.7	6.9	1.8	**	B > D	0.08
2nd grade	10.7	2.7	10.1	2.0	9.5	2.1	8.6	1.6	**	A & B > D	0.10
3rd grade	11.5	2.0	10.9	1.9	10.5	1.4	10.6	1.6			0.03
4th grade	14.6	3.0	14.0	2.7	12.9	2.9	12.6	2.5	**	A > D	0.09
5th grade	16.8	4.5	14.8	3.5	16.7	2.1	15.4	3.4	**		0.03
6th grade	19.4	6.3	16.3	5.0	17.3	3.3	15.8	3.4	*	A > D	0.09

Note: Unit is kilogram. M = mean, SD = standard deviation.
$*p < .05$, $**p < .01$

Table 9 shows differences among schoolchildren in forward bend. No significant difference was found. The findings revealed that early-born children scored similarly to late-born children, indicating that RAE was non-existent for flexibility.

Table 9. The relative age effect (RAE) on bending forward

	Group A		Group B		Group C		Group D		Main effect	Multiple comparison	Effect size
	M	SD	M	SD	M	SD	M	SD			
1st grade	26.6	9.9	30.7	6.0	29.0	6.5	31.9	7.1			0.06
2nd grade	30.1	7.7	31.1	6.6	33.0	7.2	28.8	6.7			0.08
3rd grade	32.1	6.9	30.1	5.0	31.7	3.9	31.0	6.2			0.04
4th grade	31.1	5.0	32.3	7.1	31.3	6.5	29.4	7.1			0.01
5th grade	39.8	8.2	39.2	8.8	40.0	4.9	35.2	9.9			0.03
6th grade	36.8	9.9	37.7	9.9	39.3	8.2	35.2	9.9			0.02

Note: Unit is centimeter. M = mean, SD = standard deviation.

Table 10 shows that no significant differences existed among participating children in the quick jump. Early-born children scored similarly to late-born children: therefore: RAE was non-existent for agility.

Table 10. The relative age effect (RAE) on the quick jump

	Group A		Group B		Group C		Group D		Main effect	Multiple comparison	Effect size
	M	SD	M	SD	M	SD	M	SD			
1st grade	26.5	6.6	24.6	3.7	24.9	5.7	26.2	3.8			0.02
2nd grade	30.2	2.6	30.1	4.8	28.4	6.8	27.9	8.1			0.02
3rd grade	31.3	4.9	30.6	7.4	2934	3.5	31.3	5.0			0.01
4th grade	37.6	5.0	38.7	4.8	34.9	5.9	35.3	6.5			0.06
5th grade	43.0	6.4	43.2	5.1	43.9	6.0	43.2	7.9			0.03
6th grade	43.0	6.4	43.2	5.1	43.9	6.0	43.2	7.9			0.03

Note: Unit is the number of times. M = mean, SD = standard deviation.

Table 11 shows that no significant differences were found among children in sit-ups: early-born children scored about the same as late-born children. Therefore, no RAE was observed for muscle endurance.

Table 11. The relative age effect (RAE) on sit-ups

	Group A		Group B		Group C		Group D		Main effect	Multiple comparison	Effect size
	M	SD	M	SD	M	SD	M	SD			
1st grade	13.0	4.9	13.0	4.1	13.8	4.2	11.7	4.8			0.05
2nd grade	16.5	3.0	15.9	4.9	15.7	6.3	14.3	5.8			0.02
3rd grade	16.8	3.8	16.0	4.3	14.0	5.5	15.3	3.6			0.05
4th grade	18.1	4.4	17.1	5.9	16.4	6.1	17.3	6.8			0.01
5th grade	23.8	4.6	24.8	6.3	23.5	6.4	23.2	6.4			0.01
6th grade	18.6	6.5	16.1	5.0	18.5	6.8	18.2	5.6			0.02

Note: Unit is the number of times. M = mean, SD = standard deviation.

Table 12 shows differences based on the multistage fitness test. In 2nd grade, Group A scored significantly higher than Groups C and D. In other grades, no significant differences were found. Early-born children scored higher than late-born children did, indicating the existence of RAE for whole body endurance.

Table 12. The relative age effect (RAE) on the multistage fitness test

	Group A		Group B		Group C		Group D		Main effect	Multiple comparison	Effect size
	M	SD	M	SD	M	SD	M	SD			
1st grade	25.6	11.8	22.8	11.0	18.9	9.1	19.2	8.2			0.06
2nd grade	34.2	16.8	25.5	9.0	23.9	11.5	22.3	10.7	**	A > C & D	0.09
3rd grade	34.0	19.5	31.6	15.1	29.6	14.7	31.8	14.0			0.03
4th grade	42.8	18.4	44.5	22.0	44.1	20.7	41.9	23.9			0.03
5th grade	57.9	23.8	49.6	10.4	52.8	18.4	55.3	20.7			0.02
6th grade	52.6	21.6	51.4	17.7	53.0	17.1	50.8	15.9			0.07

Note: Unit is the number of times. M = mean, SD = standard deviation.
$^{**}p < .01$

4 Discussion

The impact of relative age on physical size and motor ability among children aged 6–12 years has been highlighted. Regarding physical size (height and weight), early-born children were more physically developed compared to late-born children. The RAE on height was observed in all grades. Thus, early-born children may have a physical advantage within the same grade.

In terms of motor ability, early-born children scored impressively in some grades compared with late-born children in the 50 m sprint, standing broad jump, softball throw, assessment for grip strength, forward bend, and multistage fitness test. Specifically, RAE was characteristically observed in fundamental powers such as speed, jumping power, throwing power, and grip strength. However, for flexibility, agility, muscle endurance, and whole body endurance, it was not typically observed. These results suggest that RAE is likely to appear in terms of motor abilities, which directly reflect physical size. On the other hand, it is not likely to exist for those motor abilities that require training or practice over a long period. Thus, among children of elementary school age, early-born children may have the advantage in motor abilities over late-born children.

In children aged 4–6 years, RAE was observed in physical size [20, 21]. Thus, the findings of this study were consistent, indicating that the influence of RAE on physical size continues from at least 4 to 12 years. RAE was limited in motor abilities—speed and throwing power—in children aged 4–6 years. The findings of this study were also consistent. However, in children aged 7–12 years, RAE was observed in more kinds of motor abilities—jumping power, grip strength, and whole body endurance. This suggests that RAE in physical aspects became evident in elementary school age.

Regarding index of effect size, RAE's influence was low to middle level and was likely to decrease as children reached the upper grade. This finding suggests that differences in physical size based on birth month decrease with age. This observation may explain why RAE has generally been regarded as insignificant. However, the number of successful experiences in physical activities during childhood may be skewed. Thus, we have to pay much attention to the fact that early born children may have more opportunities to succeed.

The influence of RAE appears in physical education achievements and school sports representation. Early-born students generally receive better evaluations for physical education activities and opportunities to represent their schools in sports activities [26, 27]. Greater speed, power, strength, height, and weight contribute to better performance in sports. Therefore, children with the advantage in these areas are more likely to be identified as "outstanding" or "gifted." Additionally, they may have the advantages in selection by scouts and/or coaches for representation in sports. In Japan, although RAE on physical education attainment has not been clarified yet, this phenomenon may cause inequalities.

Teachers and coaches should consider RAE when planning physical education activities for children. Early-born children are profiting from an initial relative age advantage and are likely to be perceived as the most talented in their respective age groups because RAE is not understood, as Musch and Grondin [3] also suggested. Accordingly, increased understanding of RAE among teachers may lead to the provision of age-appropriate support for children's healthy development and fair achievement across all grade levels.

The findings of this study indicate that RAE influences physical aspects of sports participation among Japanese elementary schoolchildren, implying in some cases that early-born children may have an advantage over late-born children in terms of sports achievements.

5 Conclusion

The influence of RAE was observed in the physical size and motor abilities of Japanese elementary schoolchildren. Thus, we propose the need to acknowledge the existence of this phenomenon in school education systems and among teachers and coaches who could benefit from this awareness and its relation to physical education and sports.

References

1. Merton, R.K.: The Matthew effect in science. Science **159**, 56–63 (1968)
2. Rigney, D.: The Matthew Effect: How Advantage Begets Further Advantage. Columbia University Press, New York (2010)
3. Musch, J., Grondin, S.: Unequal competition as an impediment to personal development: a review of the relative age effect in sport. Dev. Rev. **21**(2), 147–167 (2001)
4. Grondin, S., Deshaies, P., Nault, L.: Trimestres de naissance et participation au hockey et au volleyball. La Revue Québecoise de l'Activité Physique **2**(3), 97–103 (1984). (in French)

5. Barnsley, R.H., Thompson, A.H., Legault, P.: Family planning: football style. The relative age effect in football. Int. Rev. Sociol. Sport **27**, 77–86 (1992)
6. Romann, M., Fuchslocher, J.: Influence of the selection level, age and playing position on relative age effects in Swiss women's soccer. Talent Dev. Excell. **3**(2), 239–247 (2011)
7. Thompson, A.H., Barnsley, R.H., Stebelsky, G.: "Born to Play Ball": the relative age effect and major league baseball. Sociol. Sport J. **8**, 146–151 (1991)
8. Delorme, N., Raspaud, M.: The relative age effect in young French basketball players: a study on the whole population. Scand. J. Med. Sci. Sports **19**(2), 235–242 (2009)
9. Nakata, H., Sakamoto, K.: Relative age effect in Japanese male athletes. Percept. Motor Skill **113**(2), 570–574 (2011)
10. Delorme, N., Raspaud, M.: Is there an influence of relative age on participation in non-physical sports activities? The example of shooting sports. J. Sports Sci. **27**, 1159–1166 (2009)
11. Bisanz, J., Morrison, F.J., Dunn, M.: Effects of age and schooling on the acquisition of elementary quantitative skills. Dev. Psychol. **31**(2), 221 (1995)
12. Morrison, F.J., Smith, L., Dow-Ehrensberger, M.: Education and cognitive development: a natural experiment. Dev. Psychol. **31**(5), 789 (1995)
13. Bell, J.F., Daniel, S.: Are summer-born children disadvantaged? The birthdate effect in education. Oxford Rev. Educ. **16**, 67–80 (1990)
14. Dickinson, D.J., Larson, J.D.: Effects of chronological age in months on school achievement. J. Educ. Res. **56**, 492–493 (1963)
15. Hauck, A.L., Finch, A.: The effect of relative age on achievement in middle school. Psychol. Schools **30**(1), 74–79 (1993)
16. Caroll, H.C.M.: Season of birth and school attendance. Br. J. Educ. Psychol. **62**, 391–396 (1992)
17. Diamond, G.H.: The birthdate effect—a maturational effect? J. Learn. Disabil. **16**(3), 161–164 (1983)
18. Maddux, C.D.: First-grade entry age in a sample of children labeled learning disabled. Learn. Disabil. Q. **3**(2), 79–83 (1980)
19. Davis, B.G.: Does age of entrance affect school achievement? Elem. School J. **80**(3), 133–143 (1980)
20. Kawata, Y., Iizima, M., Izutsu, S., Yamada, K., Kamimura, A., Hirosawa, M.: The relative age effect on physical fitness, motor ability, physical competence and evaluation from kindergarten teachers among Japanese young children. In: Proceedings of the 8th Asian Preschool Physical Education Conference and PE Workshop for Kindergarten Teachers, China, CD-ROM, Beijing (2012). 5 pages
21. Kawata, Y., Kamimura, A., Izutsu, S., Yamada, K., Iijima, M., Hirosawa, M.: Does the relative age effect exist in physical size, motor ability, awareness of physical activity, and kindergarten teachers' evaluation of young Japanese children? In: Abstract of the ISSP 13th World Congress of Sports Psychology, p. 247 (2013)
22. Kawata, Y., Kamimura, A., Izutsu, S., Yamada, K., Iizima, M., Mizuno, N., Hirosawa, M.: Relative age effect in physical and psychological development in young Japanese children and associated problems for kindergarten teachers. In: Ahram, T., Karwowski, W., Marek, T. (eds.) Proceedings of the 5th International Conference on Applied Human Factors and Ergonomics, AHFE 2014, pp. 414–419. AHFE Conference, Poland (2014)
23. Kamimura, A., Kawata, Y., Hirosawa, M.: The relationship between birth month, physical size, motor ability and physical activity evaluated by kindergarten teachers among Japanese young children. Juntendo Med. J. **62**(Suppl. 1), 104–108 (2016)

24. Kawata, Y., Kamimura, A., Oki, K., Yamada, K., Hirosawa, M.: Relative age effect on psychological factors related to sports participation among Japanese elementary school children. In: Salmon, P., Macquet, A.-C. (eds.) Advances in Human Factors in Sports and Outdoor Recreation, pp. 199–211. Springer International Publishing, Switzerland (2016)
25. Cohen, J.: Statistical Power Analysis for the Behavioral Science. Academic Press, New York (1969)
26. Bell, J.F., Massey, A., Dexter, T.: Birthdate and ratings of sporting achievement: analysis of physical education GCSE results. Phys. Educ. Sport Pedagogy **2**, 160–166 (1997)
27. Cobley, S., Baker, J., Wattie, N., McKenna, J.: Annual age-grouping and athlete development. Sport Med. **39**(3), 235–256 (2009)

Non-apparatus, Quasi-apparatus and Simulations Tests in Diagnosis Positive Health and Survival Abilities

Roman Maciej Kalina(✉) and Władysław Jagiełło

Department of Combat Sports, Faculty of Physical Education,
Gdansk University of Physical Education and Sport, Gdansk, Poland
kom.kalina@op.pl, wjagiello1@wp.pl

Abstract. Many of the recommendations relating to the estimated of positive health (also by WHO experts) emphasizes the need to measure the following somatic health indicators: BMI, resting HR, systolic and diastolic heart pressure, aerobic capacity, muscle strength, flexibility, balance. Ignored is the need to measure the mental health, social health. Innovative is a comprehensive approach that takes into account the two stages. First – the subjective sense of various positive health indices covers three dimensions: somatic A, mental B, social C and D dimension, which represents sense of indices and assessment reflecting individual's survival abilities. The second – the measurement of these indicators using mainly non-apparatus, quasi-apparatus and simulations tests. Simple tools measuring positive health and survival abilities can ensure high reliability of the diagnosis. The necessary condition is to meet methodological criteria.

Keywords: Mental health · Social health · Somatic health · Specific motoric tests · Verbal simulation tests

1 Introduction

Practitioners (physiotherapists, internists, neurologists, occupational physicians, coaches, physical education teachers, etc.) apreciate simple diagnostic tools. The necessary condition is to meet methodological criteria. ***Non-apparatus test*** is defined as motoric test (exercise endurance test) of the required reliability (accurate and reliable), which use does not require even the simplest instruments, whereas ***quasi-apparatus test*** – can be conducted with simple instruments (a stopwatch, a ruler, a measuring tape, etc.) [1]. Simulations tests (imitate reality representing it in a simplified fashion) include two categories: motoric (self-defense, loss of balance and fall, safe fall, etc. [2]) and mental (pictures tests, verbal simulation etc. [3]).

Many of the recommendations relating to the estimated of positive health (also by WHO experts) emphasizes the need to measure the following somatic health indicators: BMI, resting HR, systolic and diastolic heart pressure, aerobic capacity, muscle strength, flexibility, balance [4–7]. *Positive health* – a concept of health related to the quality of life and to the capability possessed by an individual. This term relates more to the development, than to the simple coping skills. In the physiological context it may be

T. Ahram (ed.), *Advances in Human Factors in Sports, Injury Prevention and Outdoor Recreation*, Advances in Intelligent Systems and Computing 603,
DOI 10.1007/978-3-319-60822-8_12

perceived as a state which is characterized by: (a) the absence of the disease; (b) low level of the severity of risk factors of the civilization diseases; (c) an adequate capacity of adaptive mechanisms responsible for the control of the external environment, the physical effort in particular [8].

Innovative approached considers a comprehensive paradigm which takes into account the two stages. First – the subjective sense of various positive health indices covers three dimensions: somatic A, mental B, social C and D dimension, which represents sense of indices and assessment reflecting individual's survival abilities. The second – the measurement of these indicators using mainly non-apparatus, quasi-apparatus and simulations tests

The aim of this study is recommendation of the Sense of Positive Health and Survival Abilities (SPHSA) questioner [9] as a simple tool diagnosing declared and empirically verified profile of these phenomena based on non-apparatus, quasi-apparatus and simulations tests.

2 Contents of SPHSA Questioner

The profile based on the subjective sense of various positive health indices covers three dimensions: somatic A, mental B, social C and D dimension, which represents sense of indices and assessment reflecting individual's survival abilities. The sense of intensity of particular indices is evaluated in the 1 to 5 scale (where its value is as follows: 1 very low, 2 low, 3 average, 4 high, 5 very high). The "0" index is used for the purpose of evaluation of specific abilities (D dimension). The arithmetic mean of indices (after decomposition to diagnostic values) calculated for particular dimensions (from A to D) constitutes a general measure of a given health dimension and survival abilities. The

Table 1. Way of the disintegration (of transforming and documenting) of indicators declared to diagnostic values [2].

Dimension	Indicator	Evaluated index (X/O)					
		0	1	2	3	4	5
A somatic health	BMI				X		→5
				2 ←		X	
			1 ←				X
	Resting HR		X				→5
				X		→4	
				2 ←		X	
			1 ←				X
	Systolic blood pressure				X		→5
				2 ←		X	
			1 ←				X
	Diastolic blood pressure				X		→5
				2 ←		X	
			1 ←				X
B mental health	Aggressiveness		X				→5
				X		→4	
				2 ←		X	
			1 ←				X
	Sense of fear		X				→5
				X		→4	
				2 ←		X	
			1 ←				X

arithmetic mean calculated for A to D indices represents the most general index of SPHSA [9]. Part of declared indicators somatic (A) and mental health (B) subjects transforming and documenting according to accepted criteria and evaluation (Table 1)

Table 2. Way of using questionnaire SPHSA (subjective assessment) [2].

Profile of Sense of Positive Health and Survival Abilities indices

/X subjectively estimated indicator ○ empirically diagnosed indicator/

female student among active on daily basis with the highest SPHA indicator/

.... AWF Katowice 21.10.2011

Surname first name or pseudonym (code) place date

age [year] 23 height [cm] 168 weight [kg] 58 sex: M (F)

physical activity: occasionally [ ] daily basis [X] sport(s) snowboard acrobatics (sport), volleyball

estimated value of the index [contractual points]:

(0) lack of ability (1) very low (2) low (3) average (4) high (5) very high

Dimension	Indicator	Evaluated index [X/○] 0	1	2	3	4	5
A somatic health	BMI				X	→	5
	Resting HR				X		
	Systolic blood pressure				X	→	5
	Diastolic blood pressure				X	→	5
	Aerobic capacity					X	
	Anaerobic capacity					X	
	Muscle strength					X	
	Flexibility					X	
	the arithmetic mean of points:	X 4,25			○......		
B mental health	Aggressiveness			X	→	4	
	Sense of fear		X	→	→	→	5
	Stress coping skills					X	
	Tolerance					X	
	the arithmetic mean of points:	X 4,333			○......		
C social health	Respecting „fair play" rule					X	
	Respecting supreme values						X
	Responsibility					X	
	the arithmetic mean of points:	X 4,25			○......		
D survival abilities	Body balance disturbance tolerance skills						X
	Precision skills before and during activity						X
	Safe falling skills					X	
	Self-defence skills					X	
	Swimming ability					X	
	Lifesaving skills in water				X		
	First aid skills					X	
	Survival abilities in solitude					X	
	the arithmetic mean of points:	X 4,125			○......		

General positive health profile and survival abilities [X/○ arithmetic means indicators]

	Dimension	0,5	1	1,5	2	2,5	3	3,5	4	4,5
A	Somatic health								X	
B	Mental health									X
C	Social health								X	
D	Survival abilities								X	

Sense of Positive Health and Survival Abilities: SPHSA [arithmetic mean A+B+C+D]	X 4,24	○......

Profile based on declarations of tested (Table 2) subjects to empirical verification. Never contrariwise. With some exceptions, declared profile can be verified by using mostly non-apparatus, quasi-apparatus and simulations tests.

2.1 Somatic Health

For measuring blood pressure, height and body mass, professional tools which fulfills technological criteria are used. For assessing aerobic and anaerobic capacity, as well as some trials of muscle strength usage of a stopwatch is sufficient for measuring a time of physical effort (quasi-apparatus tests).

During usage of tables of recommended tests (e.g. International Committee on the Standardisation of Physical Fitness Test [10], EUROFIT [11]) it is necessary to use measuring tape to evaluate results of trials, measured from the point of reference.

However, for measuring of flexibility, using of one's own fingers is enough (Fig. 1) [1], whereas own feet's could be used to measure distance of long jump (the strength of legs muscles).

a b c

Fig. 1. Example of measured flexibility: (a) results below the determined line; (b) initiation of detailed measurement from dactylion III to determined line; (c) continuation of the detailed measurement (in this example the raw score '0.5–' indicates the relatively high level of flexibility) [1].

Methodology of adjusting recommended motor tests to criteria of evaluation SPHSA is freely available [12]. Dariusz Mosler [13, 14] empirically verified usability of presented above non-apparatus flexibility test. Recommendations of tested adults who participated in health-related training based on judo indicates this test among other verified flexibility tests.

2.2 Survival Abilities

Tests for diagnosing this variable remains in strong relations with test with A dimension (somatic health). In a group of specific motor tests the *precision skills before and during activity* – PSBDA (Index D_2) consists of 10 tasks. Within the framework PSBDA aerobic capacity (Index A_5) is measured by tasks 3, 5, 7, 9 constitute 4 × 30 s Burpee Test (intensive effort, during each 30 s divided by a 60-second break, during which the precision skills are measured); the sum of repetitions is an indicator of aerobic capacity (Index A_5). An incomplete cycle is documented as 0 [12]. Anaerobic capacity (Index A_6): the results of task 3 is at the same time an indicator of anaerobic capacity (Index A_6). If only anaerobic capacity is determined under certain conditions (options), the use of 60 s Burpee Test while testing males is justified [12].

Body balance disturbation tolerance skills (Index D_1) is measured using results of "Rotational Test" [15] in to possible versions: non-apparatus test (only points); quasi-apparatus test (additionally measuring time of a test).

Safe falling skills (Index D_3) evaluated by test for safe falls (verbal and points evaluation) [16]. In some way, an alternative test can be used: *the susceptibility test of the body injuries during the fall* – STBIDF [17].

Self-defense skills (Index D_4) are measured by recommended *basic self-defense skills test* – BSDST [18]. Alternatively, result of *testing fights in a vertical posture* – TFVP [19, 20] could be used. This is an example of an adaptation of fun form of martial arts [21] for assessing *self-defense skills.*

3 Empirical Argumentation: Somatic Health (A)

Pilot application of SPHSA shows many apparent convergences of results (e.g. identical Index A and not even one concordance of declared and empirically verified scores) and significant discrepancies of Index A (difference 0.375) with e.g. 38% of compatibility of detailed (Table 3).

Table 3. Declared and empirically verified profiles SPHSA two physiotherapy female students, 23-year old [12].

declares occasional physical activity

Dimension	Indicator		Evaluated index [×/O] 0	1	2	3	4	5
A somatic health	A_1	BMI				×	O	
	A_2	Resting HR			O	×		
	A_3	Systolic blood pressure				×	O	
	A_4	Diastolic blood pressure				×	O	
	A_5	Aerobic capacity			O	×		
	A_6	Anaerobic capacity			O	×		
	A_7	Muscle strength				×	O	
	A_8	Flexibility			O	×		
	the arithmetic mean of points:		× 3.00			O 3.00		

declares daily physical activity and long-years swimming experience

Dimension	Indicator	Evaluated index [×/O] 0	1	2	3	4	5
A somatic health	A_1					⊗	
	A_2					×	O
	A_3					O	×
	A_4					O	×
	A_5						⊗
	A_6					O	×
	A_7					⊗	
	A_8					O	×
			× 4.625			O 4.250	

4 Empirical Argumentation: Index SPHSA

So far, there are 5 published studies [9, 22–25] informing about results of SPHSA questionnaire (declared profile). Students who declares higher daily physical activity are characterized by higher SPHSA. Ukrainian female students not only do not declare increased physical activity, but also have lower subjective SPHSA (Table 4).

Table 4. Declared Index SPHSA Polish and Ukrainian students (the average and standard deviation)

Subject [references]	Gender	N	Occasionally physical activity	N	Physically active on daily basis
Physiotherapy Polish [9]	Female	100	**3.479** ± 0.38	22	**3.641** ± 0.34
Tourism and recreation Polish [21]	Female	34	**3.427** ± 0.364	24	**3.74** ± 0.27
Physical education Polish [22]	Female	21	**3.757** ± 0.256	31	**4.080** ± 0.21
Paramedic sciences Polish [23]	114 females, 69 males	119	**3.70** ± 0.42	64	**3.74** ± 0.44
Paramedic sciences Ukrainian [24]	Female	467	**2.90** ± 0.36		

5 Methodological Remarks and Recommendations

Scientific description of criteria of empirical evaluation of SPHSA [12] fulfills high methodological standards. Recommended motor tests based on validation studies verified multiple times by researchers from many polish and foreign facilities. SPHSA questionnaire subjects to open formula in a meaning of possibility of using most of tests from health dimensions A and B, adapted to age, gender, physical abilities etc. It is necessary to adopt indicators to defined criteria of evaluation [12]. For example in an assumption, that for elder people (and at higher risk) one of most difficult challenges is balance loss and fall, there is one basic question: do I have sufficient muscle strength and flexibility for optimal amortization of a fall and stand up independently? For such diagnosis there is simple *non-apparatus safe falls preparations test* (N-ASFPT) [26]. Tests for evaluation of mental health (pictures tests, verbal simulation etc. [3]) are currently adopting for evaluation of social health and will be used in SPHSA questionnaire. They availability will not be widespread because of necessity of medical secrecy.

6 Conclusion

Pilot studies using a SPHSA questionnaire (first stage) and comparison with results of motor tests (dimension A and D) and verbal simulation tests (size B and C) confirms the high diagnostic value of this method. Many references to a papers written by authors of this publication is an effect of over twenty years of research projects, which completely fill a of currently developing *prophylactic and therapeutic agonology* [27–29].

References

1. Kalina, R.M.: Applying non-apparatus and quasi-apparatus tests in a widely understood concept of health promotion – an example of flexibility measurement and assessment. Arch. Budo **8**(3), 125–132 (2012)
2. Jagiełło, W., Wójcicki, Z., Barczyński, B.J., Litwiniuk, A., Kalina, R.M.: Optimal body balance disturbance tolerance skills as a methodological basis for selection of the firefighters to solve difficult tasks of rescue. Ann. Agric. Environ. Med. **21**(1), 148–155 (2014)
3. Klimczak, J., Barczyński, B.J., Podstawski, R., Kalina, R.M.: The level of bravery and aggressiveness of the sports activity organisers for the youth – simulation research. Arch. Budo **12**, 345–354 (2016)
4. Sherrington, C., Michaleff, Z.A., Fairhall, N., Paul, S.S., Tiedemann, A., Whitney, J., Cumming, R.G., Herbert, R.D., Close, J.C.T., Lord, S.R.: Exercise to prevent falls in older adults: an updated systematic review and meta-analysis. Br. J. Sports. Med., 1–10 (2016). doi: 10.1136/bjsports-2016-096547
5. Paffenbarger Jr., R.S., Olsen, E.: Lifefit. An Effective Exercise Program for Optimal Health and Longer Life. Human Kinetics, Champaign (1996)
6. Chakravarthy, M.V., Joyner, M.J., Booth, F.W.: An obligation for primary care physicians to prescribe physical activity sedentary patients to reduce the risk of chronic health conditions. Mayo Clin. Proc. **77**, 165–173 (2002)
7. American College of Sport Medicine (ACMS): ACM's Guidelines for Exercise Testing and Prescription, 7th edn. Lippinncott Wiliams & Wilkins, Philadelphia (2006)
8. Kuński, H.: Trening zdrowotny osób dorosłych. Poradnik lekarza i trenera. Agencja Wydawnicza Medsportpress, Warszawa (2002). (in Polish)
9. Kalina, R.M.: The profile of sense of positive health and survival abilities indices (subjective assessment) as a diagnostic tool used in health-related training. Arch. Budo **8**(3), 179–188 (2012)
10. Pilicz, S., Przewęda, R., Dobosz, J.: Physical Fitness Score Tables of Polish Youth. Criteria for Measuring Aerobic Capacity by the Cooper Test. Studia i Monografie AWF, Warszawa (2002)
11. Council of Europe: EUROFIT, Strasburg (1993)
12. Kalina, R.M., Kalina, A.: Methods for measurement of somatic health and survival abilities in the framework of the SPHSA questionnaire – methodological aspects. Arch. Budo Sci. Martial Art Extreme Sport **9**, 17–30 (2014)
13. Mosler, D.: Usability of non-apparatus and quasi apparatus flexibility tests based on self-perception participants in health-related judo training. Arch. Budo Sci. Martial Art Extreme Sport **11**, 189–197 (2015)
14. Mosler, D.: Validity and reliability of non-apparatus and quasi apparatus flexibility tests – verification during health-related training based on judo. Arch. Budo Sci. Martial Art Extreme Sport **11**, 123–133 (2015)
15. Kalina, R.M., Jagiełło, W., Barczyński, B.J.: The method to evaluate the body balance disturbation tolerance skills – validation procedure of the "Rotational Test". Arch. Budo **9**(1), 59–69 (2013)
16. Kalina, R.M., Barczyński, B.J., Jagiełło, W., Przeździecki, B., Kruszewski, A., Harasymowicz, J., Syska, J., Szamotulska, K.: Teaching of safe falling as most effective element of personal injury prevention in people regardless of gender, age and type of body build – the use of advanced information technologies to monitor the effects of education. Arch. Budo **4**, 82–90 (2008)

17. Kalina, R.M., Barczyński, B.J., Klukowski, K., Langfort, J., Gasienica-Walczak, B.: The method to evaluate the susceptibility to injuries during the fall – validation procedure of the specific motor test. Arch. Budo **7**, 203–216 (2011)
18. Kalina, R.M., Jagiełło, W., Wiktorek, P.: Motor competence in self-defence of students of a detectives' school during their course of studies. Arch. Budo **3**(3), 1–6 (2007)
19. Kalina, R.M., Jagiełło, W., Chodała, A.: The result of "testing fights in a vertical posture" as a criterion of talent for combat sports and self-defence – secondary validation (part I: the reliability). Arch. Budo Sci. Martial Art Extreme Sport **11**, 229–238 (2015)
20. Kalina, R.M., Jagiełło, W., Chodała, A.: The result of "testing fights in a vertical posture" as a criterion of talent for combat sports and self-defence – secondary validation (part II: the accuracy). Arch. Budo Sci. Martial Art Extreme Sport **12**, 229–238 (2016)
21. Jagiełło, W., Kalina, R.M., Klimczak, J., Ananczenko, K.V., Ashkinazi, S., Kalina, A.: Fun forms of martial arts in positive enhancement of all dimensions of health and survival abilities. In: Kalina, R.M. (ed.) Proceedings of the 1st World Congress on Health and Martial Arts in Interdisciplinary Approach, HMA 2015, Czestochowa, Poland, Warsaw, 17–19 September 2015 (2015). Archives of Budo 2015: 32–39
22. Jagiełło, W., Sawczyn, S., Jagiełło, M.: The subjective profile of positive health and survival abilities in women differing as to physical activity. Arch. Budo **8**(4), 219–224 (2012)
23. Jagiełło, W., Sawczyn, S., Jagiełło, M., Michałowska-Sawczyn, M.: The subjective profile of positive health and survival abilities in female students differing as to physical activity. Teoriâ i Praktika Fizičeskoj Kul'tury **5**, 15–18 (2013)
24. Bergier, B.: The diversity of the profiles involving the sense of positive health and survival abilities of Polish students of paramedical sciences. Arch. Budo **11**, 17–25 (2015)
25. Tsos, A., Oliynyk, O.V., Szepeluk, A.: Subjective health profiles among Ukrainian students of medical vocational school. Health Prob. Civilization **10**(1), 32–38 (2016)
26. Kalina, R.M.: Non-apparatus safe falls preparations test (N-ASFPT) – validation procedure. Arch. Budo **4**, 255–265 (2013)
27. Kalina, R.M.: Agonology as a deeply esoteric science – an introduction to martial arts therapy on a global scale. Procedia Manuf. **3**, 1195–1202 (2015)
28. Kalina, R.M.: Cognitive and application barriers to the use of "Agonology in Preventive and Therapeutic Dimension". In: Salmon, P., Macquet, A.-C. (eds.) Advances in Human Factors in Sports and Outdoor Recreation, Proceedings of the HFE 2016 International Conference on Human Factors in Sports and Outdoor Recreation. Advances in Intelligent Systems and Computing, vol. 496, pp. 25–35. Springer (2016)
29. Kalina, R.M.: Innovative agonology as a synonym for prophylactic and therapeutic agonology – the final impulse. Arch. Budo **12**, 329–344 (2016)

Combined Effects of Lower Limb Muscle Fatigue and Decision Making to the Knee Joint During Cutting Maneuvers Based on Two Different Position-Sense-Acuity Groups

Xingda Qu[1(✉)] and Xingyu Chen[2]

[1] Institute of Human Factors and Ergonomics, Shenzhen University, Shenzhen, China
quxd@szu.edu.cn
[2] College of Management, Shenzhen University, Shenzhen, China

Abstract. Research concerning the combined effect of fatigue and decision making was still insufficient to fully understand the ACL injury mechanism. A controlled laboratory experiment was conducted in which 14 volunteered healthy collegiate males participated and were instructed to perform jumping followed by cutting maneuver under four simulated circumstances, i.e., pre-fatigue anticipated, pre-fatigue unanticipated, post-fatigue anticipated, and post-fatigue unanticipated. Significant increase of knee extension, valgus, and internal rotation moment as well as peak proximal tibia anterior shear force were observed after fatigue. However, only increased knee extension moment and peak proximal tibial anterior shear force were found when the task was activating by unanticipated stimulus. Knee extension and internal rotation moments together with tibial anterior shear force were magnified under post-fatigue unanticipated condition. These findings suggested that interaction effect between fatigue and decision making may restrain the sensory sensitivity of peripheral movement receptors surrounding lower extremity, resulting in decreased proprioceptive capability.

Keywords: ACL injury · Fatigue · Decision making · Position-Sense-Acuity

1 Introduction

Non-contact anterior cruciate ligament (ACL) injuries are commonly occurred during high-demand athletic activities [1, 2]. The annual incidence rates for professional athletes (0.15%–3.7%) are extremely higher than the ones for national population (0.05%), according to a systematic review [3]. Undoubtedly, ACL injury brings huge negative impacts on physiological movement ability and psychological adjustment capability to the athletes, due to long-term and rigorous rehabilitation processes as well as sharp reduced musculoskeletal tolerance. Hence, scientific research focusing on epidemiological pathogenesis of ACL injury and the contribution of those potential risk factors plays such a crucial role in providing some useful information about diagnostic methods and preventive measures to the clinicians.

T. Ahram (ed.), *Advances in Human Factors in Sports, Injury Prevention and Outdoor Recreation*, Advances in Intelligent Systems and Computing 603,
DOI 10.1007/978-3-319-60822-8_13

Muscle fatigue was suggested to be a latent risk factor for ACL injury. Generally, fatigued muscle cannot absorb the same amount of energy as when it functions well and will rapidly reach the limited degree of extension. Chappell et al. calculated knee joint angles and resultant forces/moments through gathering video graphic images and force plate data of 20 recreational athletes prior to and after finishing fatigue exercise during forward, vertical and backward stop-jump task [4]. They obtained increased peak proximal tibial anterior shear forces, increased valgus moments, and decreased knee flexion angles after subjects receiving fatigue protocol for all three tasks in spite of gender bias. The authors speculated that fatigue would alter the motor control strategies for both males and females and such modification might increase the risk of ACL rupture. Recently, Zebis et al. tested 14 team handball female players to inquire the effect of induced muscle fatigue on neuromuscular strategy within a functional side cutting movement by means of tracking the quantity of maximal isometric voluntary contraction. Since decrease of quadriceps and hamstring muscle strength was captured during the task, the authors then predicated that acute fatigue could bring a certain degree of impairs to neuromuscular activity of ACL agonist muscles [5].

Decision making also attracts more attention and interest of clinical researchers, since it is also purported as a risk factor that may improve the possibility of being subjected to ACL injury. During sport-related maneuvers, such as landing, cutting, and pivoting task, lower extremity muscles need to be activated orderly and coordinately for the sake of maintaining joint stabilization through neuromuscular control system. Deficits in neuromuscular control, which refer to improper or even disabled activation of dynamic restraints surrounding a joint after sensing external stimuli, are proved to place athletes in a dangerous situation to experience ACL rupture. The process of detecting biomechanics of action and generating corresponding reactive movement is somehow unconscious, but very essential for human beings so as to protect joints from severe damage, particularly when the disturbance is under unpredicted or sudden arrival. Houck et al. concluded that anticipation would significantly affect hip abduction angles and knee moments after calculating thorax-pelvis-hip kinematic variables together with hip and knee moments when subjects performing two anticipated and unanticipated tasks [6]. What's more, McLean et al. obtained significant increase of peak knee abduction moments during unanticipated landings (51.25 (7.41) Nm) compared to anticipated ones (38.93 (9.32) Nm) by bilaterally recording muscle pre-motor time and knee biomechanics data of 20 NCAA female athletes [7].

In general, the capability that a person can detect relative position of neighboring segment and required strength during a motion should be attributed to the proprioceptive sense. This kind of sensation is accomplished by transferring sensory information collected from different movement receptors in the joint, capsule, ligament, skin and muscle into the cortex. Then the brain starts to analyze information and convert into a synthesized picture of body's position, based on the impulses emitted by those receptors. Under ordinary circumstances, proprioceptive sense is differentiated from each individual. Poor proprioception is inherently relevant to increased postural sway, decreased balance and increased risk of knee functional decline. In order to quantify different levels of personal proprioception, researchers created a terminology, which is denoted as the position-sense-acuity (PSA), to measure the sensory power of a person.

In spite of the fact that a large number of studies have corroborated the correlation between isolated effect of muscle fatigue or decision making and the number of injured ACL cases, the combined effects among these two risk factors were still unclear. Moreover, questions like whether the extent of impact caused by muscle fatigue and decision making on people with different PSA is in the same level and whether there is some interaction effect existing between muscle fatigue/decision making factor and PSA factor, were also unknown because no previous literature contacts with such topics. Therefore, investigating these issues will dramatically deepen our understanding of ACL injury mechanism in order to design more scientific and rational intervention skills. With these facts in mind, the authors propose a series of hypothesis expressed as following:

(a) All kinematic parameters will become larger under post-fatigue or unanticipated condition compared with pre-fatigue or anticipated condition.
(b) All kinematic parameters will become the worst ones under post-fatigue with unanticipated condition.
(c) Fatigue will be the dominant factor to promote occurrence of ACL injury for both low-PSA group and high-PSA group.

2 Methods

2.1 Participants

To date, most studies relevant to ACL injury have concentrated the target population on female subjects, which appears to be unquestionable on account of usual argument. However, outcomes gained from only manipulating those vulnerable subjects cannot expose the whole comprehensive principles hiding behind the superficial phenomenon. Therefore, 14 volunteered participants who had been recruited into this project were all healthy collegiate males. Their anthropometric statistics including age, body mass and stature (denoted in the form of mean value ± standard deviation) were 23.786 ± 1.424, 68.86 ± 9.69 (kg), and 175.79 ± 6.97 (cm), respectively. They all self-reported to have no related experience or relevant training in terms of jumping-skill developing, attention/cognition fostering etc. and no previous surgical history or any type of musculoskeletal disorders. Before initiating the experiment, each subject should carefully read and voluntarily sign in an informed consent form in case of fully understanding the procedures and conditions of the project. The written informed consent was approved by the local ethics committee. During the course of doing experiment, all subjects were needed to wear formfitting shorts and sports shoes, so that they could perform the task with enough comfort.

2.2 Experimental Design

At the beginning, participants' PSA on dominant foot side was assessed by using an isokinetic dynamometer (Biodex Medical System, Inc., New York, USA). Participants were blindfolded and asked to actively match a target position (135° flexion) from the reference position (90° flexion) and hold the posture for 5 s when recording the deviation. If mean deviation was equal to or greater than 4°, the participant would belong to low PSA group. Otherwise, they would be classified into high PSA group [8, 9]. After that, eighteen reflective markers were attached on left and right side skin surface of the participant. Specifically, those locations included left and right ASIS, bilateral thigh, bilateral lateral and medial epicondyle, bilateral shank, bilateral lateral and medial malleolus, along with bilateral heel and 5th metatarsal toe shown in Fig. 1. This marker placement can help establish a lower extremity kinematic model, consisting of nine skeletal segments, i.e. bilateral foot, talus, shank, thigh and pelvis. A motion capture system (Motion Analysis Corporation, Santa Rosa, USA) was used to capture the kinematic data during the experiment.

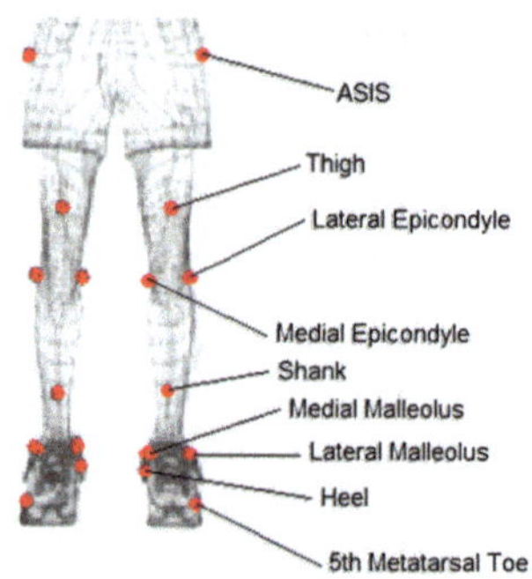

Fig. 1. Locations of eighteen reflective skin markers

Participants were required to stand on the platform with all reflective skin markers adhering on those predetermined positions accurately. After detecting the signal light, participants should immediately jump from the platform and directly land on the force plates (AMTI, USA) in front of the platform (Fig. 2). Then, participants would aggressively turn to the left/right side guided by the activation of signal light 0/1. Lower limb fatigue was induced by performing several groups of sprints combined with squats [10] without any intermediate rest. Participants were notified to do double-leg squats for 20 times immediately after running on a treadmill (Biodex Medical System, Inc., New York, USA) with 1.5 times of their ordinary running speed for 2 min. The fatigue protocol would continue to strengthen fatigue accumulation by repeating the same procedure until self-reported fatigue level reaching 18 or above from the 6–20 Borg Scale [11].

A successful trial was determined by subject's two feet landing on the force plate simultaneously and each foot stepping on only one force plate. Comprised by eight high-speed (200 fps) cameras, the motion capture system was used to collect three-dimensional coordinates of each interested segment. In addition, two force plates were also selected to record ground reaction force when subject landing on the track. The signal light actuation sequence employed in the experiment was generated by utilizing a random function based on computer programming language and was only

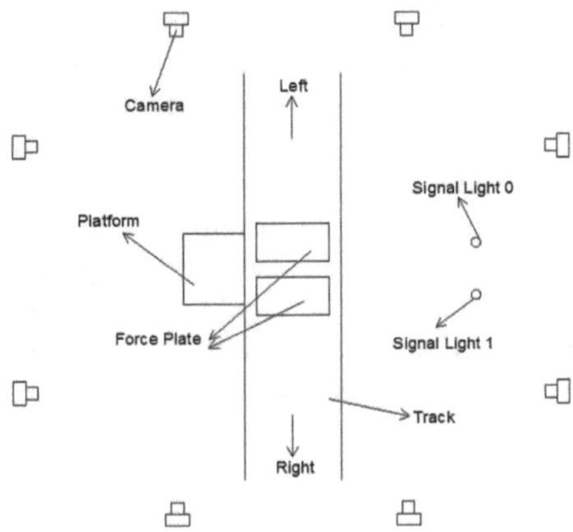

Fig. 2. Laboratory layout of the experiment

allowed to be seen by the experimenter himself. Here, number '0' represented the activation of left signal light 0, while number '1' indicated the activation of right signal light 1.

2.3 Dependent Variables

Retrieved from the objective of the project, there were three independent variables or so-called interested factors: (a) PSA, (b) fatigue, and (c) decision-making waiting to be investigated, where each of them had two distinct levels, i.e. low/high, pre/post and anticipated/unanticipated, respectively. As extremely complicated organizational structures were embedded in the knee joint, four dependent variables which had been verified to be greatly related to ACL injury, i.e. (dv1) knee flexion/extension moment, (dv2) knee varus/valgus moment, (dv3) knee internal/external rotation moment, (dv4) peak proximal tibial anterior shear force, were chosen to deeply elaborate those concerned issues.

2.4 Analysis

Through recording 3D coordinates of the reflective makers and collecting the ground reaction force, intersegmental forces and moments at both ankle joint and knee joint could be calculated by virtue of inverse dynamics approach. Since the authors were interested in the cutting period, initial time point for kinematic analysis would be defined as the moment when participant's supporting foot was just towards the direction of turning. Then, the time of termination point was designated as the moment when ground reaction force arriving at maximum value. The entire cutting period was considered to be a uniformly accelerated process owing to very short time interval. Dempster's research data [12] would be applied to estimate mass, center of mass and moment of inertia for each segment of the lower limb during the whole calculation procedure. With the aid of MATLAB software, all computed results were normalized to a 100 kg person based on every participant's body mass and were going to be imported to subsequent testing and assessment. For the sake of appraising both the individual effects of each explanatory variable and their interaction effects between any two of them, a 2^3 factorial ANOVA analysis (significance level $\alpha = 0.05$) was seemed to be appropriate and sufficient to have an insight into the complex problem, where a run was represented by a series of lower case letters (e.g. a, bc, abc). If a letter was present, the corresponding factor would be set at level 2 in that specific trial. While the absence of certain letter should be translated into the situation that the corresponding factor would be set at level 1 on that occasion.

3 Results

Results of the factorial ANOVA analysis for checking both main effects of each independent variable and interaction effects between any two of them were depicted in Tables 1 and 2. For the first dependent variable, i.e. knee flexion/extension moment (dv1), it was quite obvious that individual effects of PSA ($P < 0.001$), fatigue ($P < 0.003$) and decision-making ($P < 0.01$) were statistically significant. Besides, two interaction effects: PSA combined with fatigue ($P < 0.003$) and fatigue combined with decision-making ($P < 0.005$) also presented statistical significance. However, interaction effect between PSA and decision-making exhibited approaching statistical significance for the reason that observed P-value (0.071) was less than 0.10.

Table 1. Main effects: Mean (SD)

Dependent variables	Low-PSA	High-PSA	P-value	Pre-fatigue	Post-fatigue	P-value	Anticipated	Unanticipated	P-value
dv1 (Nm)	77.6 (106.6)	4.5 (89.8)	<0.001*	12.84 (48.28)	69.3 (134.9)	0.001*	18.61 (68.56)	63.6 (128.2)	0.008*
dv2 (Nm)	2.179 (1.706)	2.35 (2.088)	0.624	1.619 (1.054)	2.91 (2.307)	<0.001*	1.921 (1.494)	2.608 (2.194)	0.051#
dv3 (Nm)	43.52 (64.27)	64.5 (85.2)	0.122	37.5 (71.42)	70.5 (77.2)	0.016*	45.1 (76.4)	62.9 (74.9)	0.189
dv4 (N)	10.246 (3.568)	8.856 (1.383)	0.001*	8.923 (1.579)	10.179 (3.511)	0.004*	8.654 (1.911)	10.448 (3.215)	<0.001*

Notation: *indicates statistical significance (p-value < 0.05); while #indicates approaching statistical significance (p-value < 0.10).

Surprisingly, as for knee varus/valgus moment (dv2), only main effect of fatigue could be described as statistical significance ($P < 0.001$). The other two individual effects of PSA and decision-making along with all interaction effects were not regarded as significant difference from the statistical point of view. However, observed P-value for decision-making factor ($P = 0.051$) was very close to the significance level ($\alpha = 0.05$), which meant this effect might change into remarkable discrepancy in case of containing enough participants. With regard to knee internal/external rotation moment (dv3), only individual effect of fatigue was statistically significant with 0.016 P-value. Unlike the outcomes received from knee varus/valgus moment, two interaction effects,

Table 2. Combined effects: Mean (SD)

Dependent variables	Low-PSA/ Pre-fatigue	Low-PSA/ Post-fatigue	High-PSA/ Pre-fatigue	High-PSA/ Post-fatigue	P-value
dv1 (Nm)	21.22 (11.57)	134.1 (128.2)	4.5 (66.9)	4.6 (109.4)	0.001*
dv2 (Nm)	1.5 (0.666)	2.858 (2.13)	1.738 (1.34)	2.962 (2.514)	0.847
dv3 (Nm)	8.515 (5.072)	78.5 (76.5)	66.5 (92.9)	62.4 (78.5)	0.007*
dv4 (N)	8.769 (1.696)	11.723 (4.307)	9.077 (1.471)	8.635 (1.28)	<0.001*
Dependent variables	Low-PSA/ Anticipated	Low-PSA/ Unanticipated	High-PSA/ Anticipated	High-PSA/ Unanticipated	P-value
dv1 (Nm)	40.1 (46.11)	115.2 (134.7)	−2.9 (80.6)	11.9 (99.1)	0.071#
dv2 (Nm)	2.027 (1.719)	2.331 (1.713)	1.815 (1.255)	2.885 (2.593)	0.274
dv3 (Nm)	23.95 (46.76)	63.1 (73.8)	66.3 (93.7)	62.6 (77.5)	0.114
dv4 (N)	8.554 (2.37)	11.938 (3.794)	8.754 (1.346)	8.958 (1.438)	<0.001*
Dependent variables	Pre-fatigue/ Anticipated	Pre-fatigue/ Unanticipated	Post-fatigue/ Anticipated	Post-fatigue/ Unanticipated	P-value
dv1 (Nm)	14.89 (40.72)	10.8 (55.6)	22.3 (88.9)	116.3 (156.9)	0.004*
dv2 (Nm)	1.519 (0.746)	1.719 (1.3)	2.323 (1.913)	3.496 (2.547)	0.165
dv3 (Nm)	48.9 (97.6)	26.14 (24.81)	41.38 (48.42)	99.6 (89.7)	0.003*
dv4 (N)	8.535 (1.385)	9.312 (1.69)	8.773 (2.345)	11.585 (3.942)	0.018*

Notation: *indicates statistical significance (p-value < 0.05); while #indicates approaching statistical significance (p-value < 0.10).

i.e. PSA with fatigue ($P < 0.01$) and fatigue with decision-making ($P < 0.005$), appeared to be statistically significant. Interestingly, both main effects of PSA, decision-making and their interaction effect had no statistical distinctiveness.

As far as peak proximal tibial anterior shear force (dv4), it manifested that all main effects of PSA ($P < 0.003$), fatigue ($P < 0.005$), and decision-making ($P < 0.001$) together with all interaction effects among these three explanatory variables were statistically significant. Interestingly, even though knee flexion/extension moment and peak proximal tibial anterior shear force were calculated in the same sagittal plane, insignificant interaction effect between PSA and decision-making factor with respect to former response variable (dv1) converted to be statistical discrepancy when inspecting with the later one (dv4).

4 Discussion

For knee flexion/extension moment (dv1), similar trend was found in the main effect pictures of both fatigue and decision-making factor (Fig. 3). Specifically, all subjects experienced larger knee extension moment after fatigue. Moreover, the amplitude of knee extension moment became higher for all participants when receiving unanticipated signal than anticipated one. These findings were consistent with previous research. Muscle fatigue was usually initiated by a period of maximum contractions, which pressurized muscle's nerves to generate force close to the upper limit of their ability. Once ultimate exhaustion point was reached, the frequency of nerve's signal would decrease to the bottom line, thereby diminishing the capability to control joint stability. Thomas et al. examined the effects of quadriceps and hamstrings fatigue on lower extremity neuromechanics by asking 25 male and female volunteers to do single-leg forward hops onto a force platform [13]. They measured significant increases in knee extension angle and moment at the time of peak vertical ground reaction force following fatigue. In addition, Yu et al. also captured that young female soccer players had decreased hip and knee flexion angles at initial ground contact after comparing three-dimensional videographic data for 30 male and 30 female adolescents with a stop-jump task. Fatigue of lower limb muscles might alter knee neuromechanics during dynamic activities, increasing noncontact ACL injury risk [14].

Additionally, it was quite noticeable that larger knee valgus moment (dv2) emerged at the situation of post-fatigue, regardless of the category for low/high PSA group or anticipated/unanticipated stimulus. A mass of researchers had already verified that abduction (valgus) moment at the knee joint would place ACL at risk. Hewett et al. claimed that injured ACL athletes had 2.5 times greater knee abduction moment and 20% higher ground reaction force after prospectively measuring the neuromuscular control of 205 athletes in high-risk sports, such as soccer, basketball and volleyball. Then the authors further calculated the specificity (73%) and sensitivity (78%) of using knee abduction moment to predict ACL injury [15].

Figure 3 also explicitly illustrated that subjects felt increased knee internal moment (dv3) when they finishing the fatigue protocol. Knee internal moment in the transverse plane was confirmed to be involved in the genesis of non-contact ACL injuries. Golden et al. conducted a crossover design in the laboratory to identify knee kinematics and kinetics during three running conditions with 13 female collegiate basketball athletes. They noted larger peak abduction, adduction, and internal rotation moment on the knee

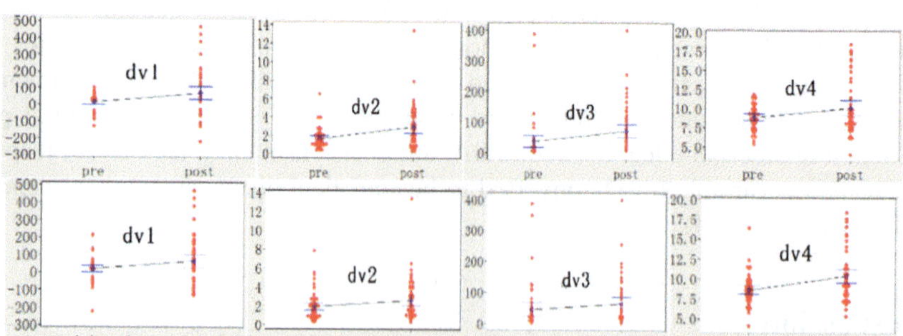

Fig. 3. Main effects of fatigue and decision-making factor for knee flexion/extension (dv1), varus/valgus (dv2), internal/external rotation (dv3) moment, and peak proximal tibial anterior shear force (dv4)

joint in a lateral false-step maneuver than normal running. The authors also pointed out that lateral false step represented a kind of mechanism for noncontact ACL injury [16].

Speaking of the peak proximal tibial anterior shear force (dv4), strong individual effects in terms of fatigue and decision-making factor were depicted in Fig. 3. Increased peak proximal tibial anterior shear force was found under the condition of being induced with fatigue and activating the task with unanticipated signal. Cadaver studies had declared that tibial anterior shear force would generate huge stress and strain to the ACL, especially when the knee was near full extension. In general, hamstrings were contracted to provide posterior tibial shear force for the sake of cutting down the force placed on the ACL. In case muscles of hamstrings could not produce adequate counterforce to balance tibial anterior shear force, the ACL would be undoubtedly at risk of sever rupture. Chappell et al. carried out an experiment to investigate knee kinetics of 10 male and 10 female recreational athletes during forward, vertical and backward stop-jump tasks. They stated that all subjects exhibited greater proximal tibial anterior shear force at the landing phase of backward stop-jump action than the others. Men instead of women had larger proximal tibial anterior shear force at the takeoff phase of both vertical and backward tasks [4]. Recently, another paper also claimed that greater peak proximal tibial anterior and lateral shear forces were recorded during landing phase of single-leg stop-jump motion [17].

The probability plot in Fig. 4 displayed knee flexion/extension moment value of two PSA groups after synthesizing the influence of both fatigue and decision-making. Such combination could be imagined as a kind of projection from three-dimensional hypothesis testing to one-dimensional analysis. Apparently, subjects in low-PSA group tended to bear knee extension moment no matter what condition they were involved with. On the contrary, people with high-PSA had an inclination to knee flexion moment during the cutting movement. It had been demonstrated that knee extension status would transmit more forces to the ACL compared to the flexion one, where the principle could be explained that if knee joint flexed adequately at the moment of landing or turning, more instantaneous load and impulsive energy would be absorbed by gastrocnemius muscle as well as achilles tendon, resulting in less impact being transferred to ligament. Current observations in this study strongly supported this point, which suggested that assessing PSA might be a simple and effective way to

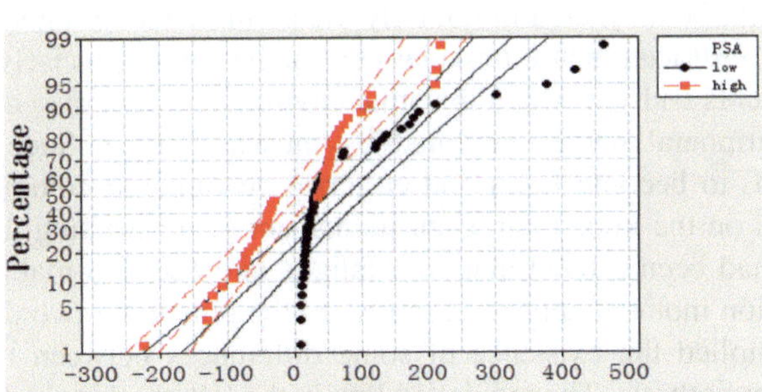

Fig. 4. Probability plot of knee flexion (<0)/extension (>0) moment under low-PSA and high-PSA group with 95% confidence interval

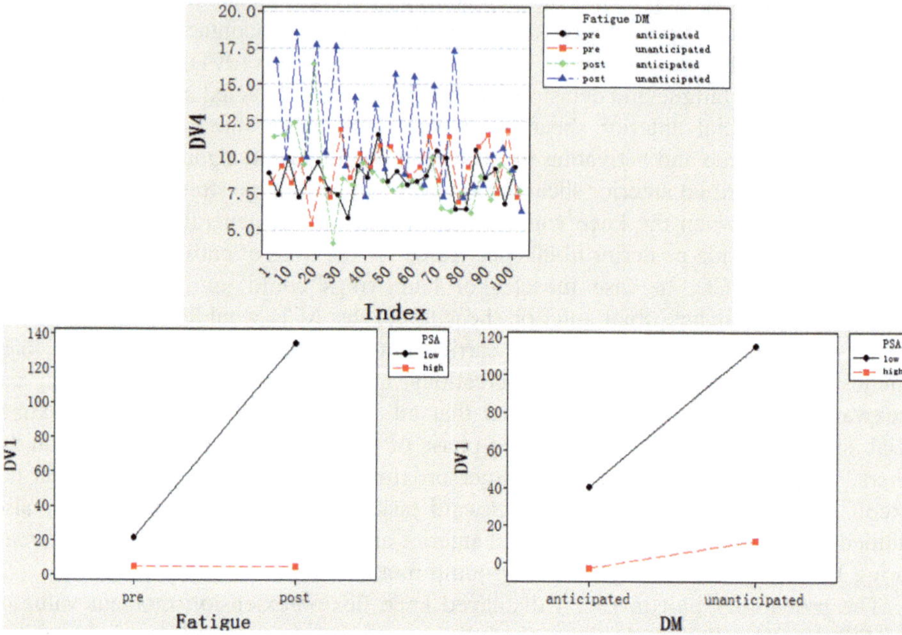

Fig. 5. Interaction effect between fatigue factor and decision-making factor for peak proximal tibial anterior shear force (upper) and interaction effects between PSA factor and fatigue as well as decision-making factor for knee flexion/extension moment (lower)

preliminarily judge whether a person was prone to ACL injury for clinical examination and sports training.

The upper picture (shown in Fig. 5) was plotted to do a contrast about peak proximal tibial anterior shear force within four subgroups (namely pre-anticipated, pre-unanticipated, post-anticipated and post-unanticipated). Obviously, the majority of blue-triangular dots were located extremely higher than all the other ones, indicating that compositive action of muscle fatigue and unanticipated signal maximally exacerbated tibial anterior shear force to be acted on the knee joint. The reason for such circumstance might be attributed to not only the perturbance brought by unanticipated irritation, which obstructed neuromuscular control system to precisely perceive spatial location of segments and joints in a dynamic environment, but also decreased sensory sensitivity of peripheral nerves inhibited by muscle fatigue. Such scenario would promote the ACL to become faster and easier to be ruptured completely, generating unexpected harm on the knee joint, even permanent or devastating damage.

Although it had been proved that both fatigue and decision-making would aggravate knee extension moment applied to the ACL, the remaining two diagrams in lower part of Fig. 5 implied the existence of some differences between them, which was beyond original hypothesis. The red dotted line in the left chart looked like a horizontal straight line, which seemingly suggested that for high-PSA group, fatigue did not have any impact on their performance. Supposing if this case was true, a huge number of previous researches would be overturned. However, it was imprudent and cursory to make an inference so early. Because there were two problems during the experiment, that the authors were not so sure. The first one was that whether all participants honestly reported their fatigue level after accomplishing several sets of fatigue protocol. Erroneously evaluating the feeling of fatigue or unconsciously overstating the fatigue level at that time caused by the purpose of self-protection was actually understandable. Besides, the intermission between each trial could not be ignored, since it provided participants with some recovery time to alleviate the extent of fatigue from maximum level. But one thing, at least, was very positive that even though the above concerns were indeed happened and included in the experimental data during later analysis, people with high PSA were good at adjusting their own physical conditions to accommodate themselves to the surrounding changes and were more quickly to recover from fatigue than those with low PSA.

Despite the statistical power, these two charts tried to convey the message that high-PSA group was more impressionable to the interference of unanticipated incentive rather than muscle fatigue. Conversely, it was not decision-making hazard but fatigue factor that dominated core position to affect the motion of low-PSA group. Although such explanation seemed to be plausible, this was the first time to discuss impact of ACL risk factors by taking the PSA into account. These findings highlighted in this project would be served as the impetus for encouraging extensive ongoing researches to aim at excavating a more precise understanding of noncontact ACL injury mechanism in the future. However, relatively small sample size together with slightly unequal fatigue level induced to every participant was the limitation of this study. Therefore, future work should focus on not only incorporating more subjects into the experiment, but exploring scientific and feasible intervention methods to enhance both proprioceptive and neuromuscular capabilities of some certain groups during prevention strategy training programs.

Acknowledgements. This work was supported in part by the Guangdong Provincial Department of Science and Technology (Project #: 2014A020220014), and the Natural Science Foundation of Guangdong Province (Project #: 2015A030313553). The author would like to appreciate Mr. Nan Wang for his assistance in data collection and draft preparation, and Dr. Xingyu Chen for her assistance in manuscript editing.

References

1. Waldén, M., Hägglund, M., Magnusson, H., Ekstrand, J.: Anterior cruciate ligament injury in elite football: a prospective three-cohort study. Knee Surg. Sports Traumatol. Arthrosc. **19**, 11–19 (2011)
2. Warme, W.J., Feagin, J.A.: Ski injury statistics, 1982 to 1993, Jackson Hole ski resort. Am. J. Sports Med. **23**, 597–600 (1995)
3. Moses, B., Orchard, J., Orchard, J.: Systematic review: annual incidence of ACL injury and surgery in various populations. Res. Sports Med. **20**, 157–179 (2012)

4. Chappell, J.D., Herman, D.C., Knight, B.S., Kirkendall, D.T., Garrett, W.E., Yu, B.: Effect of fatigue on knee kinetics and kinematics in stop-jump tasks. Am. J. Sports Med. **33**, 1022–1029 (2005)
5. Zebis, M.K., Bencke, J., Andersen, L.L., Alkjær, T., Suetta, C., Mortensen, P.: Acute fatigue impairs neuromuscular activity of anterior cruciate ligament-agonist muscles in female team handball players. Scand. J. Med. Sci. Sports **21**, 833–840 (2011)
6. Houck, J.R., Duncan, A., DeHaven, K.E.: Comparison of frontal plane trunk kinematics and hip and knee moments during anticipated and unanticipated walking and side step cutting tasks. Gait Posture **24**, 314–322 (2006)
7. McLean, S.G., Su, A., van den Bogert, A.J.: Development and validation of a 3-D model to predict knee joint loading during dynamic movement. J. Biomech. Eng. Trans. ASME **125**, 864–874 (2003)
8. Ribeiro, F., Mota, J., Oliveira, J.: Effect of exercise-induced fatigue on position sense of the knee in the elderly. Eur. J. Appl. Physiol. **99**, 379–385 (2007)
9. Singh, N.B., Nussbaum, M.A., Madigan, M.A.: Evaluation of circumferential pressure as an intervention to mitigate postural instability induced by localized muscle fatigue at the ankle. Int. J. Ind. Ergon. **39**, 821–827 (2009)
10. Borotikar, B.S., Newcomer, R., Koppes, R., McLean, S.G.: Combined effects of fatigue and decision making on female lower limb landing postures: central and peripheral contributions to ACL injury risk. Clin. Biomech. **23**, 81–92 (2008)
11. Ribeiro, F., Santos, F., Goncalves, P., Oliveira, J.: Effects of volleyball match-induced fatigue on knee joint position sense. Eur. J. Sport Sci. **8**, 397–402 (2008)
12. Dempster, W.T.: Space requirements of the seated operator. Aerospace Medical Research Laboratory WADC Technical Report 55-159 (1955)
13. Thomas, A.C., McLean, S.G., Palmieri-Smith, R.M.: Quadriceps and hamstrings fatigue alters hip and knee mechanics. J. Appl. Biomech. **26**, 159–170 (2010)
14. Yu, B., McClure, S.B., Onate, J.A., Guskiewicz, K.M., Kirkendall, D.T., Garrett, W.E.: Age and gender effects on lower extremity kinematics of youth soccer players in a stop-jump task. Am. J. Sports Med. **33**, 1356–1364 (2005)
15. Hewett, T.E., Myer, G.D., Ford, K.R., Heidt Jr., R.S., Colosimo, A.J., McLean, S.G.: Biomechanical measures of neuromuscular control and valgus loading of the knee predict anterior cruciate ligament injury risk in female athletes: a prospective study. Am. J. Sports Med. **33**, 492–501 (2005)
16. Golden, G.M., Pavol, M.J., Hoffman, M.A.: Knee joint kinematics and kinetics during a lateral false-step maneuver. J. Athl. Train. **44**, 503–510 (2009)
17. Wang, L.I.: The lower extremity biomechanics of single- and double-leg stop-jump tasks. J. Sports Sci. Med. **10**, 151–156 (2011)

Activation Sequence Patterns of Forearm Muscles for Driving a Power Wheelchair

Chi-Wen Lung[1,2,4], Chien-Liang Chen[5], Yih-Kuen Jan[1,2,3], Li-Feng Chao[4], Wen-Feng Chen[6], and Ben-Yi Liau[6(✉)]

[1] Rehabilitation Engineering Lab, University of Illinois at Urbana-Champaign, Champaign, IL, USA
cwlung@asia.edu.tw
[2] Kinesiology and Community Health, University of Illinois at Urbana-Champaign, Champaign, IL, USA
[3] Computational Science and Engineering, University of Illinois at Urbana-Champaign, Champaign, IL, USA
[4] Creative Product Design, Asia University, Taichung, Taiwan
[5] Department of Physical Therapy, I-Shou University, Kaohsiung, Taiwan
[6] Biomedical Engineering, Hungkuang University, Taichung, Taiwan
byliau@hk.edu.tw

Abstract. The purpose of this study was to investigate muscle activities of the upper limbs while driving a power wheelchair. Eleven healthy individuals were recruited to perform four joystick control tasks, including forward, backward, left-turn, and right-turn. The results of this study would establish a norm for evaluating the controls of a power wheelchair in children with cerebral palsy. The surface electromyographic monitor (EMG) was used to record the contractions of extensor carpi ulnaris (ECU), flexor carpi ulnaris (FCU), and flexor carpi radialis (FCR). The integration of EMG signals was used to quantify the muscle efforts. The results showed that participants use more muscle efforts in ECU during backward early, but during left-turn later. The results of the forearm muscle activations can be used to guide training of children with cerebral palsy to drive a power wheelchair.

Keywords: Cerebral palsy · Electromyography · Joystick · Power wheelchairs

1 Introduction

The work about children with mobility impairments using power mobility equipment in clinical practice is widely discussed [1–4]. One of the most common mobility impairments in childhood is cerebral palsy (CP). The incidence rate of CP is estimated to be about 2 to 3 in every 1,000 infants [5]. Those with severe mobility impairments especially need assistive devices for independently moving. Clinicians and researchers have investigated and promoted the use of power wheelchairs for independent mobility in children with CP. However, children with CP often have upper extremity impairments that affect controlling a power wheelchair. The upper extremity impairments could affect

T. Ahram (ed.), *Advances in Human Factors in Sports, Injury Prevention and Outdoor Recreation*, Advances in Intelligent Systems and Computing 603,
DOI 10.1007/978-3-319-60822-8_14

muscle activities in children with CP during controlling a joystick to maneuver a power wheelchair. The approach, muscle sequence analysis, is a well-established method to study controls of various activities. Muscle activation patterns are not well understood while controlling a joystick to maneuver a power wheelchair.

Muscle activation patterns during controlling a joystick of a power wheelchair in children with CP need to be improved by training process. Therefore, it is important to study muscle activities at various driving tasks while driving a power wheelchair. Although muscle contractions are complicated, the hypotheses could be muscles sequence patterns are correlated to the driving performance of joystick with power wheelchair [6, 7].

The main purpose of this study was to investigate the sequences of muscle activities of the upper limbs while driving a power wheelchair at four tasks, including forward, backward, right-turn, and left-turn. It can help to improve the training process for using a joystick to control a power wheelchair in children with CP.

2 Methods

2.1 Participants

The joystick of power wheelchair was adopted to assess the muscle activity responses of the driving performance tasks. For comparison purposes, we need to know the muscle activation pattern in normal people before the children with cerebral palsy test. There were 11 subjects recruited in this study (N = 11; mean ± SD: age, 21.0 ± 0.8 years; height, 171.2 ± 8.4 cm; weight, 64.1 ± 14.5 kg; body mass index, 21.7 ± 3.8 kg/m^2). Study protocols were reviewed by the Institutional Review Board at University.

2.2 Data Acquisition

EMG signals were acquired from wrist muscles that mainly contribute to wrist flexion-extension while controlling the joystick during driving tasks. The surface electromyographic (EMG) was used to record the contraction of three right forearm muscles, including extensor carpi ulnaris (ECU), flexor carpi ulnaris (FCU), and flexor carpi radialis (FCR). Anatomical guidelines were used for approximate locations of muscles [8]. Also, electrical stimulator was used for verifying muscle bundle location (KS-138 electronic therapeutic massager, E-neng Tech Enterprise Co., Ltd., Taiwan). Before applying the EMG electrodes (MA-300, Motion Lab Systems, Inc., Baton Rouge, LA, USA) to the skin, the surface was cleaned by alcohol for decreasing noise. The signals were amplified with a bandwidth of 20 to 450 Hz and a gain value of 1000. At the same time, serial analog X/Y signals from the power wheelchair's joystick, which reflects forward-backward and left-right joystick movements. A joystick equipped on a power wheelchair with a right-armrest (FC-100, Rong-Jan, Inc., Taiwan) was adopted in this study for driving performance task was used in this study for the driving performance tasks. The data acquisition of the joystick X/Y coordinates and EMG were synchronized recording at 1000 Hz by using the analog to digital converter (USB-6218, National Instrument, Austin, TX, USA) (Figs. 1 and 2).

Fig. 1. Schematic diagram of the experimental setup. The forearm of the participants were securely lay on the armrests of the power wheelchair for controlling the joystick in forward, backward, right-turn, and left-turn.

2.3 Driving Tasks

The driving task in this study was performed in an indoor for keeping stable experimental surrounding. The testing protocol included four tasks: forward, backward, left-turn, and right-turn. The order in which the participants tested each task was randomized. Subjects were asked for the training trials several times before starting the measurement. It helped the subjects to become familiar with joystick controlling and to get their maximum performance as possible before starting the evaluation trials. The experiment operator would inform the participants after each trial about the direction error of X/Y data of joystick and encouraged him/her to improve it. Each trial remains 5 s for moving the joystick with a 5 s rest period till next trial. Subjects would keep controlling the joystick until they were asked to stop, it resulted about 15 trials. Three stable trials were saved for data analysis.

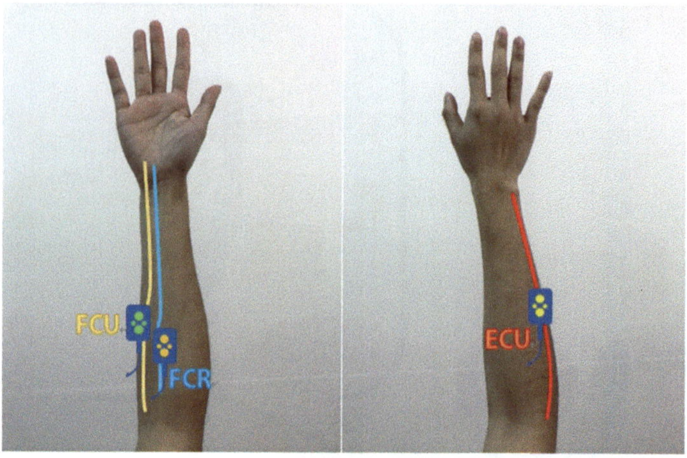

Fig. 2. The EMG electrodes placed at the specific muscles including extensor carpi ulnaris (ECU), flexor carpi ulnaris (FCU), and flexor carpi radialis (FCR).

2.4 Statistics

The individual EMG driving patterns are then time normalized, expressed as a percentage of total cycle [9]. The maximum EMG level from each specific muscle was identified during the entire set of forward driving conditions and then used to normalize the muscle's activity to this value [10]. We focused on how the EMG responses in each condition (backward, left-turn, and right-turn).

To quantify the EMG response to the driving task, the time and magnitude of maximum EMG, and the integrated EMG (iEMG) were calculated for each condition. The iEMG was calculated for the integrated value between the beginning and end of the response.

A one-way analysis of variance (ANOVA) was used to test effects of each condition (backward, left-turn, and right-turn) on the driving-related EMG. LSD post-hoc test was applied to identify differences between each condition. When statistical significance was encountered, LSD post-hoc test for multiple comparisons was performed in order to determine which muscle response was significantly different from the others using $p < 0.05$ as the criterion of statistical significance.

3 Results and Discussions

The results showed that participants use more muscle efforts (iEMG) in ECU during backward and right-turn, FCR during left-turn (Fig. 3); the participants use more EMG activation or the peak contraction value (peak EMG) in ECU during backward and right-turn, FCR during left-turn (Fig. 4); the EMG later peak contraction time (peak time

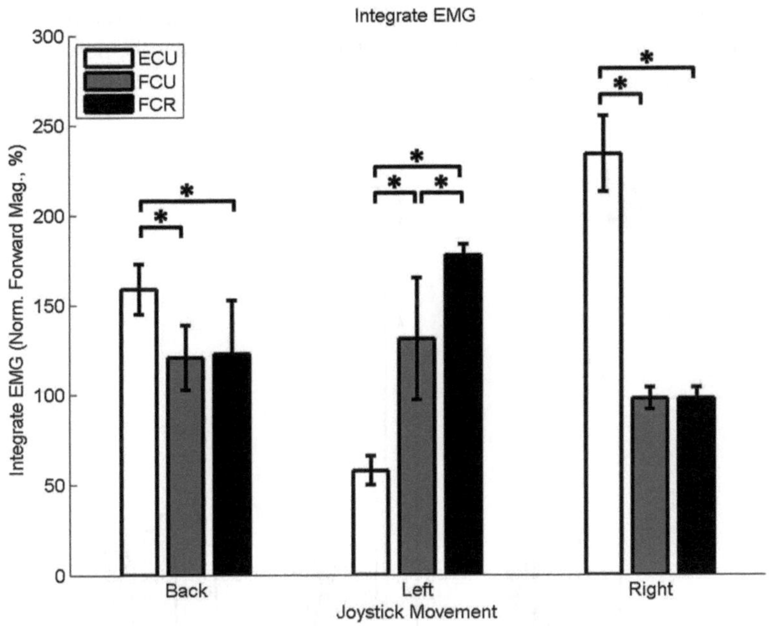

Fig. 3. Integrated EMG (iEMG). extensor carpi ulnaris (ECU), flexor carpi ulnaris (FCU), and flexor carpi radialis (FCR).

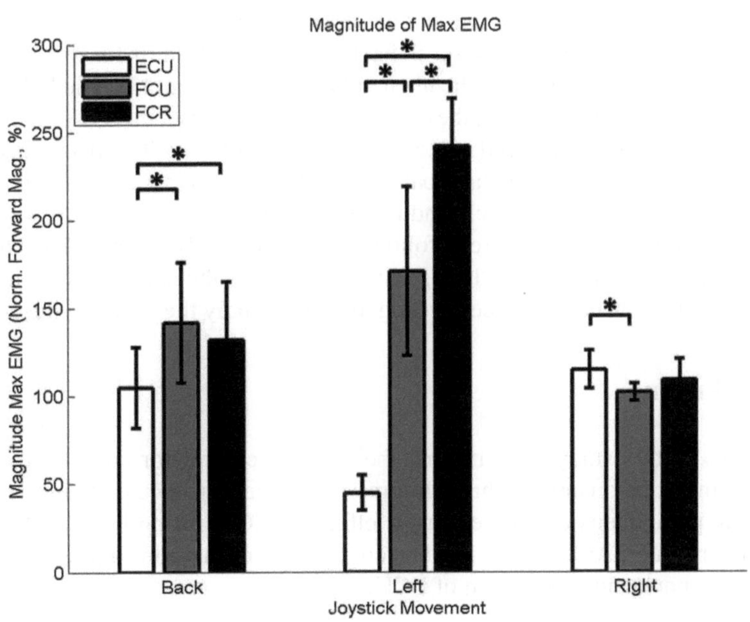

Fig. 4. Magnitude of maximum EMG (peak EMG). Extensor carpi ulnaris (ECU), flexor carpi ulnaris (FCU), and flexor carpi radialis (FCR).

EMG) was found in ECU during left-turn (Fig. 5). The resulting forearm muscle activation sequences can help clinicians improve the training process for power wheelchair joystick control in children with cerebral palsy.

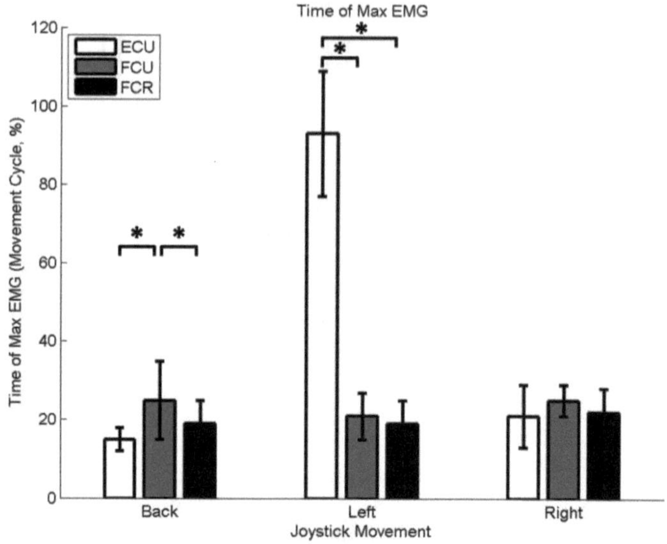

Fig. 5. Time of maximum EMG (peak time EMG). Extensor carpi ulnaris (ECU), flexor carpi ulnaris (FCU), and flexor carpi radialis (FCR).

ECU showed the major condition is backward and right-turn for driving joystick were. ECU move the hand toward the ulnar side (i.e., adduction) [7]. The ECU muscles improve wrist joint stability during the backward movement. But the left-turn movement (abduction in this study) displayed reduced the contribution of ECU. That indicates ECU interrupt at the base of the 5th metacarpal in anatomy. However, the major contribution of controlling joystick is the muscle and movement of thumb.

The limitations of this study are as follows: (1) the test walkway was flat and without any obstacle. It is not the reality living environment; (2) the subjects were not power wheelchair users. It needs advance investigation, especially the specific muscles.

4 Conclusions

The findings of this study indicate that the joystick control for controlling a power wheelchair might be quantified through forearm muscle sequence patterns. The results showed that participants use more muscle efforts in ECU during backward early, but during left-turn later. An effective joystick control for maneuvering a power wheelchair may require a powerful contraction of ECU.

Acknowledgements. We are grateful for the Ministry of Science and Technology of the Republic of China for financially supporting this research under contracts MOST 105-2221-E-468-005 and 104-2218-E-468-001.

References

1. Jones, M.A., McEwen, I.R., Hansen, L.: Use of power mobility for a young child with spinal muscular atrophy. Phys. Ther. **83**, 253–262 (2003)
2. Lynch, A., Ryu, J.C., Agrawal, S., Galloway, J.C.: Power mobility training for a 7-month-old infant with spina bifida. Pediatr. Phys. Ther. **21**, 362–368 (2009). The official publication of the Section on Pediatrics of the American Physical Therapy Association
3. Ragonesi, C.B., Chen, X., Agrawal, S., Galloway, J.C.: Power mobility and socialization in preschool: a case study of a child with cerebral palsy. Pediatr. Phys. Ther. **22**, 322–329 (2010). The official publication of the Section on Pediatrics of the American Physical Therapy Association
4. Ragonesi, C.B., Chen, X., Agrawal, S., Galloway, J.C.: Power mobility and socialization in preschool: follow-up case study of a child with cerebral palsy. Pediatr. Phys. Ther. **23**, 399–406 (2011). The official publication of the Section on Pediatrics of the American Physical Therapy Association
5. Surveillance-of-Cerebral-Palsy-in-Europe (SCPE): Surveillance of cerebral palsy in Europe: a collaboration of cerebral palsy surveys and registers. Dev. Med. Child Neurol. **42**, 816–824 (2000)
6. Lobo-Prat, J., Keemink, A.Q., Stienen, A.H., Schouten, A.C., Veltink, P.H., Koopman, B.F.: Evaluation of EMG, force and joystick as control interfaces for active arm supports. J. Neuroeng. Rehabil. **11**, 68 (2014)
7. Sangani, S.G., Raptis, H.A., Feldman, A.G.: Subthreshold corticospinal control of anticipatory actions in humans. Behav. Brain Res. **224**, 145–154 (2011)
8. Delagi, E.F., Perotto, A.: Anatomic Guide for the Electromyographer–The Limbs. Charles C Thomas Publisher, Springfield (1980)
9. Carollo, J.J., Matthews, D.: Strategies for clinical motion analysis based on functional decomposition of the gait cycle. Phys. Med. Rehabil. Clin. North Am. **13**, 949–977 (2002)
10. Ricamato, A.L., Hidler, J.M.: Quantification of the dynamic properties of EMG patterns during gait. J. Electromyogr. Kinesiol. **15**, 384–392 (2005). Official Journal of the International Society of Electrophysiological Kinesiology

Direct and Indirect Effect of Hardiness on Mental Health Among Japanese University Athletes

Shinji Yamaguchi[1,3(✉)], Yujiro Kawata[1,2], Nobuto Shibata[1,2,3], and Masataka Hirosawa[1,2]

[1] Graduate School of Health and Sports Science, Juntendo University, 1-1, Hiraga-gakuendai, Inzai-shi, Chiba 270-1695, Japan
tyaka051115@gmail.com

[2] Department of Health and Sports Science, Juntendo University, 1-1, Hiraga-gakuendai, Inzai-shi, Chiba 270-1695, Japan

[3] Institute of Health and Sports Science and Medicine, Juntendo University, 1-1, Hiraga-gakuendai, Inzai-shi, Chiba 270-1695, Japan

Abstract. This study examined direct and indirect effects of hardiness on mental health among Japanese university athletes. Participants were 760 university athletes (men = 524, women = 236, M_{age} = 19.9, SD = 1.22). The questionnaire comprised demographic information (gender, role in team), the Revised Athlete Hardiness Scale (RAHS), the Adolescent Resilience Scale (ARS) and General Health Questionnaire-30 (GHQ-30). For men, RAHS score had a direct negative effect on GHQ scores, while for women, RAHS score had no effect on GHQ score. In regular players, the RAHS score had a direct negative effect on GHQ scores, while in non-regular players, It had no effect on GHQ scores. In conclusion, hardiness directly influenced mental health among Japanese university athletes and the influence of hardiness on mental health differed by gender and team role. This suggests that hardiness may be a factor in promoting mental health among Japanese male and regular athletes.

Keywords: Hardiness · Resilience · Mental health · University athlete

1 Introduction

University athletes are exposed to many stresses because of their exposure to various competitive activities, in addition to stressors such as academic achievement and interpersonal relationships in university school settings [1]. For example, stress factors include conflict of human relationships with supervisors and teammates, excessive competition, difficulties in achieving compatibility with study, lack of pleasure, and experiences of failure [2]. Thus, researchers are exploring various viewpoints in order to flexibly cope and respond to these stress factors. Specifically, researches have been increasing focusing on the concepts of hardiness and resilience.

Hardiness is defined as an "attitude to support stress countermeasures of resilience, a concept that refers to skills [3]." If hardiness (commitment, control and challenge) is

T. Ahram (ed.), *Advances in Human Factors in Sports, Injury Prevention and Outdoor Recreation*, Advances in Intelligent Systems and Computing 603,
DOI 10.1007/978-3-319-60822-8_15

high, people can maintain mental and physical health even under stressful circumstances [4].

Resilience is a proximity concept of hardiness. It is defined as "resilience in the psychological aspect of an individual who can recover it even if the mental state falls into a negative state under difficult circumstances [5]."

Hardiness and resilience are common concepts in terms of psychological strength under adversity. As a reason for this, Nakanishi and Tamase [6] studied what role hardiness and resilience played in a stress situation. It was observed that, Hardiness and resilience directly relieved stress and this increased stress coping, affected reevaluation of cognition, and relieved stress indirectly. In addition, since hardiness has a more prominent effect in enhancing coping than resilience, improving hardiness is regarded as a basic way of treating stress [6]. According to Rahimian and Asgharnejad [7], hardiness and resilience showed a significant negative correlation with anxiety and depression. They further reported that people who were high in hardiness and resilience could deal with various adverse events. In addition, in the sports field, Karamipour, Hejazi, and Yekta [8] studied hardiness and resilience of athletes and non-athletes. They suggested that athletes were higher in hardiness and resilience than non-athletes. For these reasons, the concepts should not be evaluated a good or bad but the functions and merits of both concepts should be considered. Given this context, several papers and books have been published on these concepts, such as "Hardiness as the Pathway to Resilience [9]," "Does Hardiness Improve Resilience? [10]," and "Personal Hardiness as the Basis for Resilience [9]." Thus, under stress, hardiness should be paid more attention compared with resilience [11, 12]. Since this view is similar to that of Nakanishi and Tamase [6], it is safe to assume that hardiness has a higher impact on stress than resilience.

Therefore, in this study, we dealt with the concepts of hardiness and resilience and examined the influence of these on mental health. Moreover, based on the study of Nakanishi and Tamase [6], we assumed that hardiness is the root of human mind and we constructed a hypothesis model. The hypothetical model of this study is a model that predicts the relationship between athlete's hardiness, resilience, and mental health. We also examined the direct and indirect effect of hardiness on mental health by mediating resilience.

2 Method

2.1 Participants

We collected data from 511 Japanese university athletes (305 men, 206 women). Mean age was 19.9 ± 1.21 and age range was 18–22 years. The survey was conducted in October 2016.

2.2 Measurements

Socio-demographic Questions. We asked the students about their gender, age, grade, sports event, and competitive level. The responses were analyzed using both description and the selection formula.

Revised of Athletic Hardiness Scale (RAHS). This scale comprised 12 items of 3 factors (commitment, control, and challenge). The subjects responded to all the items of each question, and their responses were scored on a 4-point scale, ranging from "I completely disagree" (1 point) to "I completely agree" (4 points). Regarding the reversal item, the score was reversed. The average of total item scores was calculated.

The Adolescent Resilience Scale (ARS). This scale comprised 21 items of 3 factors (novelty seeking, emotional regulation, and positive future orientation). Participants responses to the items were scored on a 5-point scale, ranging from "disagree" (1 point) to "agree" (5 points).

General Health Questionnaire-30 (GHQ-30). This scale had 30 items to assess mental health. The total score was the sum of the scores obtained. We adopted the Goldberg scoring method (0–0–1–1) that assigns a score of 0 for responses 1 and 2 and a score of 1 for responses 3 and 4.

2.3 Ethical Consideration

The Research Ethics Committee of the School of Health and Sports Science, Juntendo University approved this study. Prior to the study, we obtained written informed consent from all participants. Each participant was made aware of his or her right to decline cooperation at any time without repercussions, even after consenting to participate.

2.4 Statistical Analysis

First, gender differences and competition level differences of all variables were examined.

Second, to clarify the causal relationship between hardiness and resilience and mental health, we verified the model using covariance structure analysis. Goodness of Fit Index (GFI), Adjusted Goodness of Fit Index (AGFI), Comparative Fit Index (CFI), and Root Mean Square Error of Approximation (RMSEA) were used as goodness of fit indices for the model. When a statistically significant difference was confirmed in gender differences and competition level differences, simultaneous analysis of multiple populations was used in covariance structure analysis to preside gender and the competition level. In the estimation method of analysis, the maximum likelihood method was used and the variance of each latent variable was also constrained to one, and each path from the error variable to the observation variable was constrained to one to secure the discrimination of the model. We used SPSS 22.0 and AMOS22.0 for analysis.

3 Results and Discussion

3.1 Demographic Information of Participants

Table 1 shows mean, standard deviation, and correlation coefficients of hardiness (RAHS), resilience (ARS), and mental health (GHQ-30). There was a significant negative correlation between hardiness, resilience, and mental health, respectively.

Table 1. Descriptive statistics and correlations of hardiness, resilience, and mental health

	1	2	3	4	5	6	7	8	9	Mean	SD
1. GHQ-30	–									6.6	5.14
2. RAHS	−.30**	–								35.7	5.76
3. Commitment	−.30**	.73**	–							12.1	2.52
4. Control	−.26**	.80**	.37**	–						11.4	2.46
5. Challenge	−.14**	.78**	.31**	.50**	–					12.2	2.5
6. ARS	−.29**	.52**	.34**	.47**	.40**	–				77.4	11.15
7. Positive future orientation	−.25**	.43**	.26**	.40**	.34**	.79**	–			19.7	3.66
8. Emotional regulation	−.30**	.46**	.33**	.43**	.30**	.80**	.41**	–		31.2	5.39
9. Novelty seeking	−.14**	.38**	.23**	.23**	.34**	.84**	.62**	.45**	–	26.4	4.69

**$p < .01$

Note: GHQ-30: General Health Questionnaire-30, RAHS: Revised of Athletic Hardiness Scale, ARS: The Adolescent Resilience Scale.

3.2 Multiple Group Structure Modeling

Figure 1 shows the differences in mental health with consideration of gender (men or women). For men, hardiness had a direct negative effect on mental health scores, while for women, hardiness had no effect. In men, the direct effect from hardiness to mental health was confirmed, but in women, it was not confirmed. Specifically, the direct effect pass coefficient was −.27 ($p < .05$) for men, and −.26 ($p <$ n.s.) for women.

Figure 2 shows the differences in mental health with consideration of athlete's role in their team (regular player or non-regular player). In regular players, the hardiness had a direct negative effect on mental health, while in non-regular players, hardiness had no effect on mental health. In regular players, the direct effect from hardiness to mental health was confirmed, but in non-regular players, it was not confirmed. Specifically, the direct effect pass coefficient was −.41 ($p < .01$) for regular player, and −.25 ($p <$ n.s.) for non-regular player.

Indirect effect of hardiness on mental health through resilience was .05 in men, .09 in women, .02 in regular player, and .08 in non-regular player. Both of the assumed models were significant at 0.1% level. Regarding the goodness-of-fit indicator, for gender, GFI = .95, AGFI = .88, CFI = .91, and RMSEA = .08, and for athlete's role in their team, GFI = .94, AGFI = .87, CFI = .90, and RMSEA = .08.

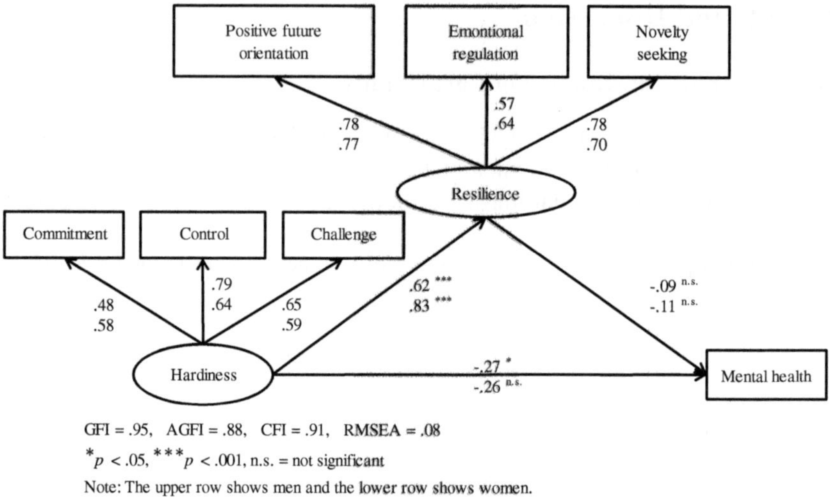

Fig. 1. The multiple group structure modeling on gender

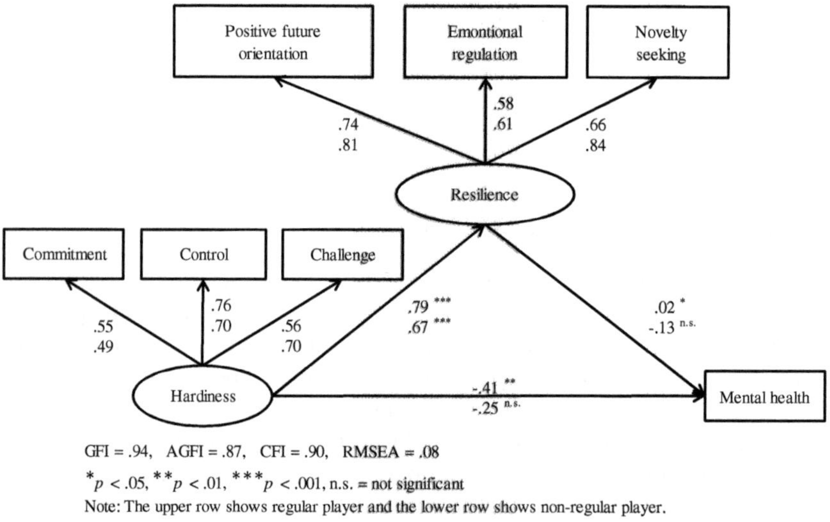

Fig. 2. The multiple group structure modeling on athlete's role in their team

4 Discussion

Regarding gender, results suggested that hardiness directly affected mental health in men, while in women, hardiness did not have the same affect. Regarding the role of team, results suggested the hardiness directly affected mental health in regular players but not in non-regular players. Regarding the direct and indirect effects examination, only the direct effect was significant. On the other hand, the direct effect of resilience had a lower impact than hardiness. Moreover, indirect effect of examination showed that the direct effect of hardiness was higher than mediation of resilience.

These results suggest that mental health is improved when hardiness is high. Hardiness had a stronger effect on mental health than resilience did. In fact, enhancing coping led to higher hardiness than resilience [6]. Thus, people who are high in hardiness actively cope with problems and perform cognitive reevaluation. Moreover, the pass coefficient of hardiness from resilience had a strong positive value. From this finding, we can conclude that enhancing hardiness also contributes to improving resilience.

The reason why the indirect effect was not significant is because there was no significant stress event between hardiness and resilience. Some kind of stress event is important because resilience is the "psychological aspect of an individual who can recover it even if the mental state falls into a negative state under difficult circumstances." On another front, hardiness can help an individual to flexibly respond to stress. For instance, combination of attitudes that provides the courage and motivation to do the hard, strategic work of turning stressful circumstances from potential disasters into growth opportunities [13]. In addition, hardiness has an adaptive effect to encourage retention of mental and physical health under stress and further enhance performance in academic and work [14]. However, a number of researches on resilience have focused on recovery from trauma experience. This means recovering from an extreme stress event (trauma experience) that cannot be dealt with hardiness is where resilience plays a more important role. Maddi and Khoshaba [15] also suggested that the "Hardiness → resilience → stress reaction" model should be adopted. Moreover, athletes have better mental health and can be more successful in their life because these two factors (hardiness and resilience) increase their perseverance when faced with difficult problems of life [8]. Therefore, we suggest that enhancing both hardiness (for overcoming stress) and resilience (for recovering from stress) contribute to improvement in mental health.

5 Conclusion

We conclude that hardiness directly influences mental health among Japanese university athletes, and that the influence of hardiness on mental health differs by gender and team role. This suggests that hardiness may be a factor in promoting mental health among Japanese male and regular athletes.

References

1. Shinji, Y., Yuki, U., Taira, S.: Developing the athletic hardiness scale for university students. Stress Manag. Res. **12**(1), 46–53 (2016)
2. Masatoshi, Y.: Variation in the motivation and burnout scale with continuation or discontinuation of sports in high school athletes. Jpn. J. Phys. Educ. Health Sport Sci. **47**(5), 427–437 (2002)
3. Miho, N., Yosuke, H., Yoshihiko, T.: The effect of self-refection on depression mediated by hardiness. Jpn. J. Psychol. **86**, 347–353 (2014)
4. Kobasa, S.C.: Stressful life events, personality and health: an inquiry into hardiness. J. Pers. Soc. Psychol. **37**, 1–11 (1979)

5. Masten, A., Best, K., Gordon, E.W.: Resilience and development: contributions from the study of children who overcome adversity. Dev. Psychopathol. **2**, 425–444 (1990)
6. Yukiko, N., Koji, T.: The role of resilience and hardiness under the stress situation. Tezukayama Univ. Bull. Psychol. **3**, 31–41 (2014)
7. Rahimian, V.E., Asgharnejad, F.A.: The relationship between psychological hardiness also ego-resiliency and mental health in adolescent and adult survivors of barn earthquake. Iran. J. Psychiatr. Clin. Psychol. **14**(1), 62–70 (2008)
8. Karamipour, M., Hejazi, M., Yekta, Z.B.: The role of resilience and hardiness in mental health of athletes and non-athletes. Bull. Environ. Pharmacol. Life Sci. **4**(2), 37–43 (2015)
9. Maddi, S.R.: Hardiness: Turning Stressful Circumstances into Resilient Growth, pp. 7–17. Springer, Dordrecht (2013)
10. Puri, R.: Does hardiness improve resilience? Int. J. Eng. Technol. Sci. Res. **3**(4), 47–49 (2016)
11. Bonanno, G.: Loss, trauma, and human resilience: how we underestimated the human capacity to thrive after extremely aversive events. Am. Psychol. **51**, 72–82 (2004)
12. Maddi, S.R.: Hardiness: an operationalization of existential courage. J. Hum. Psychol. **44**, 279–298 (2005)
13. Maddi, S.R.: Hardiness: the courage to grow from stress. J. Posit. Psychol. **18**, 160–168 (2007)
14. Maddi, S.R.: The story of hardiness: twenty years of theorizing, research, and practice. Consult. Psychol. J. Pract. Res. **54**, 173–185 (2002)
15. Maddi, S.R., Khoshaba, D.M.: Resilience at Work. AMACOM, a Division of the American Management Association, International, New York (2005)

A Real-Time Feedback Navigation System Design for Visually Impaired Swimmers

Ze En Chien(✉), Chien-Hsu Chen, Fong-Gong Wu, Nien-Pu Lin, Tong Hsieh, and Tzu Hsuan Hong

Department of Industrial Design, National Cheng Kung University, No. 1, University Road, Tainan City 701, Taiwan
lambozeen@gmail.com,
{chenhsu, fonggong}@mail.ncku.edu.tw
themaniac09@gmail.com, tony83417@gmail.com,
cc830314@gmail.com

Abstract. Due to lack of exercise, the visually impaired have a higher risk of developing critical health problems. Among all the sports, the acceptability of swimming is higher for the visually impaired. Swimming can be an excellent exercise for them, for not only maintaining good health, but also improving the illness caused by inactivity's sedentary. Due to inconvenient, environmental unfriendly, safety concern and some personal obstacle, most of the visually impaired are not willing to swim in public swimming pool. Results showed that the obstacle and unfriendly space are the main reason that affect the willingness of visually impaired to swim. A swimming goggle with real-time feedback navigation system and environmental sensor will be designed to assist, guide direction, reducing obstacles and difficulties of both swimmers and caregivers, providing a convenient way for them to take part in this activity.

Keywords: Visually impaired · Visually impaired swimmer · Swimming goggles · Goggles design

1 Introduction

About 285 million people are visually impaired worldwide. According to World Health Organization, the visual impairment case increases every year, in 285 million people, 39 million are blind and 246 have low vision [1]. Those people who have visual impairment even blindness have a higher risk of developing critical health problems due to lack of physical activity exercise [2]. The health-related fitness levels of visually impaired are generally lower than those are sighted [3].

The high risk of developing serious health due to inactivity problem of visually impaired has highlighted the importance of health fitness and taking exercises, including aerobic and anaerobic exercise, which can help them facilitate metabolism and keep healthy. Swimming can be an excellent exercise for visually impaired. Among all the sports, the acceptability of swimming is higher for visually impaired. Swimming helps to improve in range of motion, muscle strength, dynamic equilibrium, stability, locomotion and cardiovascular endurance [4]. Swimming provides various

T. Ahram (ed.), *Advances in Human Factors in Sports, Injury Prevention and Outdoor Recreation*, Advances in Intelligent Systems and Computing 603,
DOI 10.1007/978-3-319-60822-8_16

benefits for visually impaired. However, visual impaired swimmers faced some barrel that include requiring a person as a guidance, reminding them the direction and prevent them from hitting the wall while swimming [5]; safety concern of family or themselves, afraid of injury during exercising [6]; and some self-obstacles, afraid of being mock or being watched in a strange way.

This paper aims to examine and discover the difficulties and obstacles faced by the visually impaired swimmers which eventually can be solve and improve by a wearable design combining with swimming equipment.

The research hypotheses are as follows:

- **Hypothesis 1:** The visually impaired swimmer can't determine the direction of the pool and the lane of the pool.
- **Hypothesis 2:** The visually impaired swimmer feels unsafe when swimming alone during whole process.
- **Hypothesis 3:** The visually impaired swimmer knocks the wall without being notify while swimming.
- **Hypothesis 4:** The visually impaired swimmer cannot determine direction while swimming.
- **Hypothesis 5:** The wearable design increased their interest in swimming.
- **Hypothesis 6:** The visually impaired swimmer prefers independent rather than being accompany by caregiver.

The following sections will introduce the methodology and report the results of this study with an in-depth discussion and implications. The results indicated that obstacles and unfriendly space that affect the willingness of the visually impaired to swimming. This paper presents swimming goggles with real-time feedback navigation system and an environmental sensor to guide direction, reduce obstacles and difficulties, relieving the caregivers' burden and ensure the safety of visual impaired swimmers concurrently.

2 Methodology

2.1 Nonparticipant Observation Method

The participants of this study included two visually impaired swimmers, a long-term caregiver and a swimming coach recruited from Tainan City Yuming Visually Impaired Society Association. Nonparticipant observation method is to observing the completely usual swimming process of both visually impaired swimmers, including a caregiver and a swimming coach. A research method whereby the researcher observes the subjects with his or her knowledge, without taking part in the situation [7]. The researcher must keep low profile and as possible not to affect the process. The motive of nonparticipant observation method is to observing and discover the obstacles and difficulties that the visually impaired swimmers may face. The shortcomings of this method is if the observation is overt, it may lead the subject behave differently, thus affect the reliability of the data collected, the participant performance may be affect by Hawthorne effect. The participant may change their behavior improve or reduce in response to their consciousness of being observed [8, 9]. Of participants, both of them are female, both 21 years old, undergraduate student. The caregiver, 40 years old, is a mother to one of the visually impaired swimmer and the swimming coach is male, 35 years' old, had experience of teaching visually impaired swimming students.

The observation site is in the Tainan Spa World outdoor swimming pool. Before starting the observation, the researchers go thru site investigator, understand the circumstance and find a possible angle for erect photographic equipment. Two DSLR camera and one waterproof camera were used to record the whole process.

At the beginning of the observation, the participants were requested by the researcher to do what they usually do in the swimming process. In the whole process, the caregivers will accompany the visually impaired swimmers to change their swimming suit, after that the caregiver will lead the visually impaired swimmers to the swim lane. Before swimming, it will be a short warm up section. After warm up, the visually impaired swimmers started to swim in medley swimming posture (all of the four swimming posture). The caregiver and coach will stand in both starting and finishing line to remind the visually impaired swimmers before they reach the wall. The coach will whistle, if the swimmer's direction goes incline (Fig. 1).

Fig. 1. Nonparticipant observation method – actual observation while visually impaired swimmers are swimming.

After observation, researcher making unstandardized interview with the two swimmers and other participant, for example what problem that they face during swimming process. The caregiver shared her experience and knowledge of taking care visually impaired swimmers, and the risk they will faced if they swim without guidance. After understanding the swimming and guidance process, researchers manipulate into a visually impaired user to experience the difficulties during the whole swimming process.

2.2 Unstandardized Interview

The participants of this study included three visually impaired swimmers, a counselor. They were recruited from Tainan City Yuming Visually Impaired Society Association.

Unstandardized interview method was used to interview the visually impaired swimmers, including the counselor. An unstructured interview or non-directive interview is an interview in which questions are not prearranged. These non-directive interviews are considered to be the opposite of a structured interview which offers a set amount of standardized questions [10]. The form of the unstructured interview varies widely, with some questions being prepared in advance in relation to a topic that the researcher or interviewer wishes to cover. They tend to be more informal and free flowing than a structured interview, much like an everyday conversation. The disadvantages of unstandardized interview are they can only have less participants and the result most likely to stand for small sum of people, it cannot represent most of the people. Besides that, it will spend a lot of time to do the interview. Of participants, one of the visually impaired is female and the other two are male. Three of them are undergraduate student. One of them are fully blind and the other two are high-low visually impair. The counselor is male, 41 years old, work at Tainan City Yuming Visually Impaired Society Association as a full time counselor (Fig. 2).

Fig. 2. Unstandardized interview with visually impaired swimmers and the counselor.

The interview site held in the Tainan City Yuming Visually Impaired Society Association center. A recorder and a DV camera were used in the process, to record down the whole interview process. The researcher used pen and paper to record the interview.

At the beginning of the interview, the researcher first understands the situation of visually impaired and slightly get their background information. At the beginning, the interview mostly talks about their lifestyle, for example, the differences between visually impaired and sighted person, what exercise they do during their free time, what they do when they are free etc. After that, the research gradually guides the question to visually impaired swimming issue and understand what they actually do or not to do while swimming, and the obstacle and problem they face while swimming. In this whole interview, the counselor participates in it. When the visually impaired has some problem, he assisted them to answer the question and help them interpretation the problem.

3 Results

As mentioned previously, this paper aims to examine and discover the difficulties and obstacles faced by the visually impaired swimmers that eventually can be solved and improved by a wearable design combined with swimming equipment.

The results of non-participated observation and unstandardized interviews indicate that all of the hypotheses above are established. Other than that, we found the other obstacles and essential point they faced during the swimming process. The obstacles and essential point are as below: (1) the visually impaired swimmer relies on using white cane to reach swim lane and bathroom. White cane is necessary for the visually impaired, they need white cane to identify directions or objects. Without a white cane, they may felt insecurity and could be in a danger situation; (2) the visually impaired swimmers may bump into swim lane divided rope and cause injury; (3) unusual swimming posture due to unable to identify direction, the swimmers may touch the lane rope to confirm their direction. Incorrect swimming posture may reduce their performance; (4) the visually impaired swimmers may bump into other swimmers and cause serious injury.

Following the problem, there are the design specification: (1) The design should be combine with swimming equipment, and wearable device; (2) guiding function, guide the visually impaired swimmer before and after swimming (Example: go to bathroom or swim lane.); (3) bumping alert function (prevent the visually impaired swimmers from bumping into wall or other user in swimming pool); (4) direction guiding function while visually impaired swimmer is swimming.

Following the design specification, a real-time feedback navigation system has been developed in this study. This system come with a streamline look swimming goggles and a sensor, which can be fixed on the gutter, to help them recognizing the direction and prevent them from bumping into wall while swimming. Besides the swimming part, it also includes a guiding system, to guiding the user to reach the swim lane and an emergency call system, alert the safeguard when the visually impaired swimmers need help. With this design, it effectively reduces the obstacle that the visually impaired swimmers faced, thus increase their willingness of swimming to keep healthy (Fig. 3).

Fig. 3. Appearance of swimming goggles and sensor design.

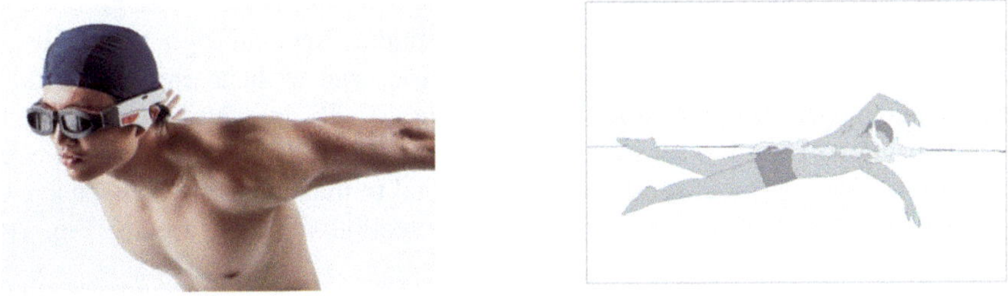

Fig. 4. Usage scenario of swimming goggles

3.1 Design Feature

The proposed real-time feedback navigation system come with two parts a swimming goggles and an environmental sensor, as shown in Fig. 4. Two environmental sensors are require in a swim lane for receive signal from swimming goggles and there is a limit of maximum 4 device users swimming in this swim lane, as shown in Fig. 5.

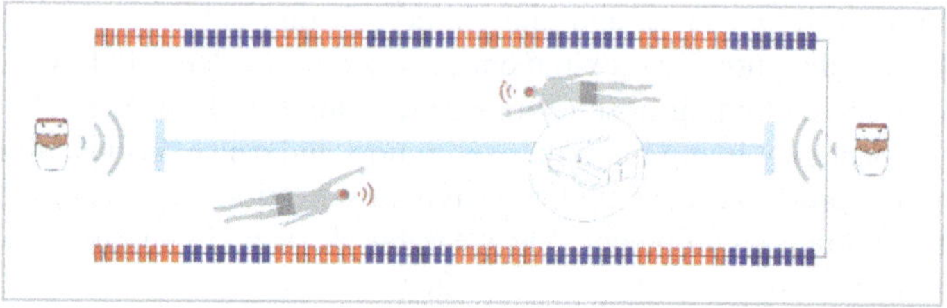

Fig. 5. User scenario: environmental sensors are place on both side of the swim lane. Both goggles of the swimmer communicate with the environmental sensors, guide direction for the swimmers.

3.2 Details of Swimming Goggles

The role of swimming goggles (see Fig. 6) is communicating with the environmental sensor for navigate, trigger emergency signal and protect the user's eye from water flushing.

Fig. 6. Swimming googles

The swimming goggles uses bone conduction method to transmit sound through the user's bone, using sound information to guide the user's direction. The bone conduct vibrator (see Fig. 7(a)) converts sound into vibrations, sent through the skull bone, directly to the user's inner ear [11]. It allows user can to hear environmental and notification sound at the same time. This help the visually impaired user reduce their fear while using it. The PPG sensor in Fig. 7(a) automatically detect abnormal oxygen concentration of the user and send emergency signal to the environmental sensor. USB port is available for charging with waterproof protection silicone cap (see Fig. 7(b)).

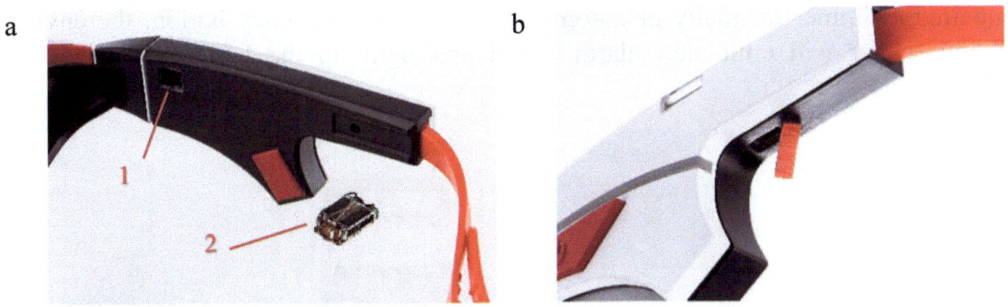

Fig. 7. Detail design of swimming goggles. (a) PPG sensor (1) and bone conduction vibrator (2) inside the swimming goggles. (b) USB port is available for charging with waterproof protection silicone cap.

The appearance shape design of swimming goggles making the user intuitively place the bone conduction and PPG sensor in the right place (see Fig. 8(a)). The raised button design with convex surface (power button and adjust volume button), increased the tactile impression of visually impaired user (see Fig. 8(b)). Adjustable silicon straps make it suitable for various head sizes (see Fig. 8(c)).

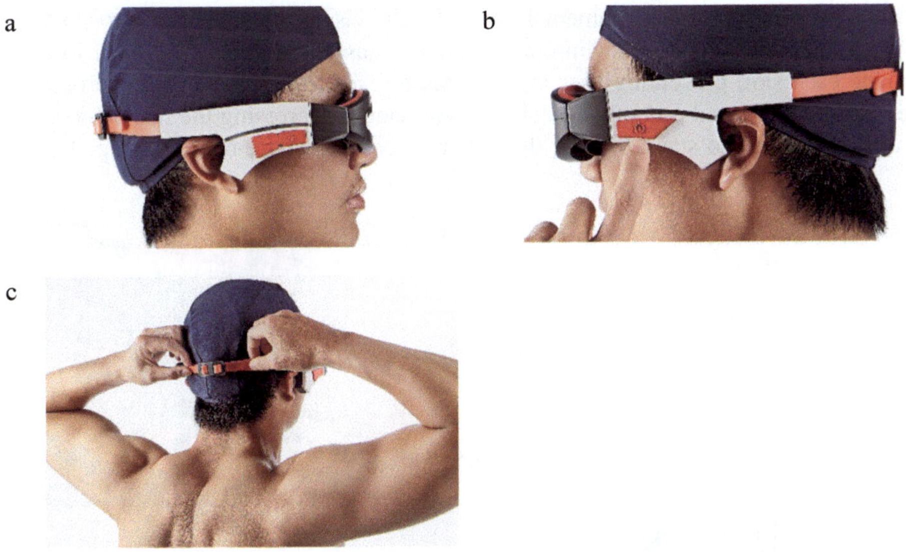

Fig. 8. Detail design of swimming goggles. (a) Shape design of swimming goggles. (b) Raised convex button design. (c) Adjustable silicon straps.

3.3 Details of Environmental Sensor

The role of environmental sensor (see Fig. 9) is to receive signal from goggles and positioning the location of the user with swimming goggles. When the visually impaired swimmer manually or automatically trigger emergency button, the environmental sensor will emit alert alarm sound and light up the LED emergency alert lighting.

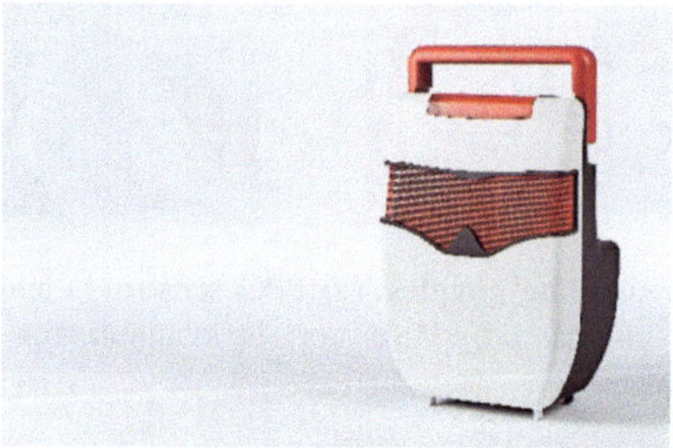

Fig. 9. Environmental sensor

The environmental sensor come with sleek margin design for comforts of touch, preventing users from accidental cut. Aside from receiving signals, the radio receiver also convey a sense of speed in appearance. When emergency occur, the goggles will send out signal to trigger emergency alarm, light up the LED on the case with distress sound to arouse the lifeguard attention (shown in Fig. 10(a)). There are three specific slots at the rear of the environmental sensor for the visually impaired user to store their white cane while they are swimming. Each slot has Braille engraved on it to differentiate each slot, preventing user from mistaking (see Fig. 10(b)). The environmental sensor will be placed on both side of the swim lane. By pressing the handle down, the case will be able to stand firm on the gutter to avoid displacement (see Fig. 10(c)).

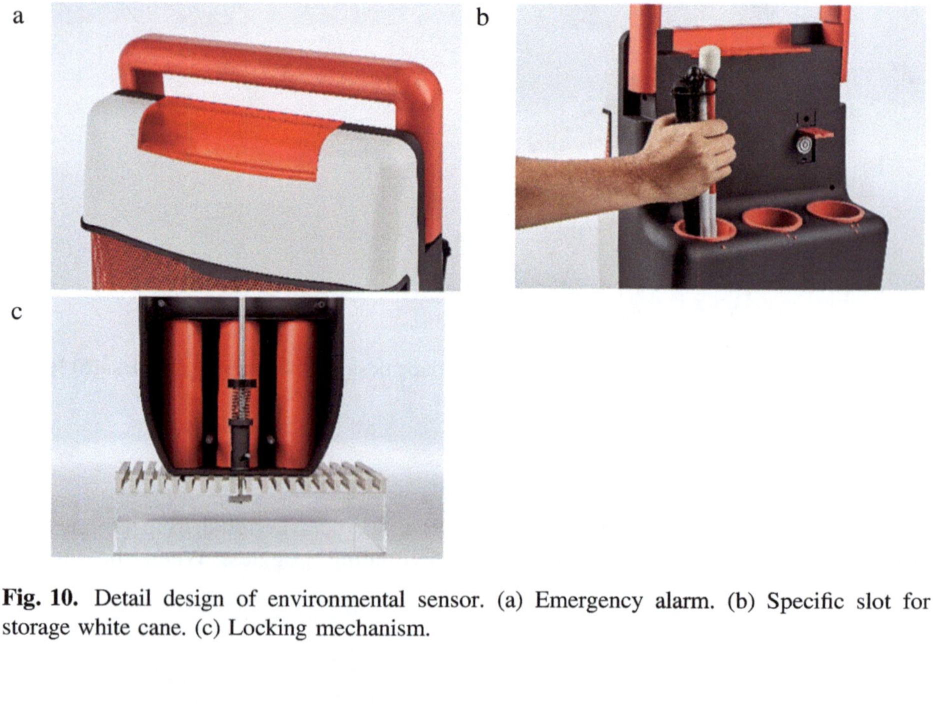

Fig. 10. Detail design of environmental sensor. (a) Emergency alarm. (b) Specific slot for storage white cane. (c) Locking mechanism.

3.4 Internal Working

By using wireless protocol to position the coordinate system, the accelerator and digital compass will be able to detect deviation and displacement.

3.5 Procedure

The complete using procedures are as below (shown in Fig. 11):

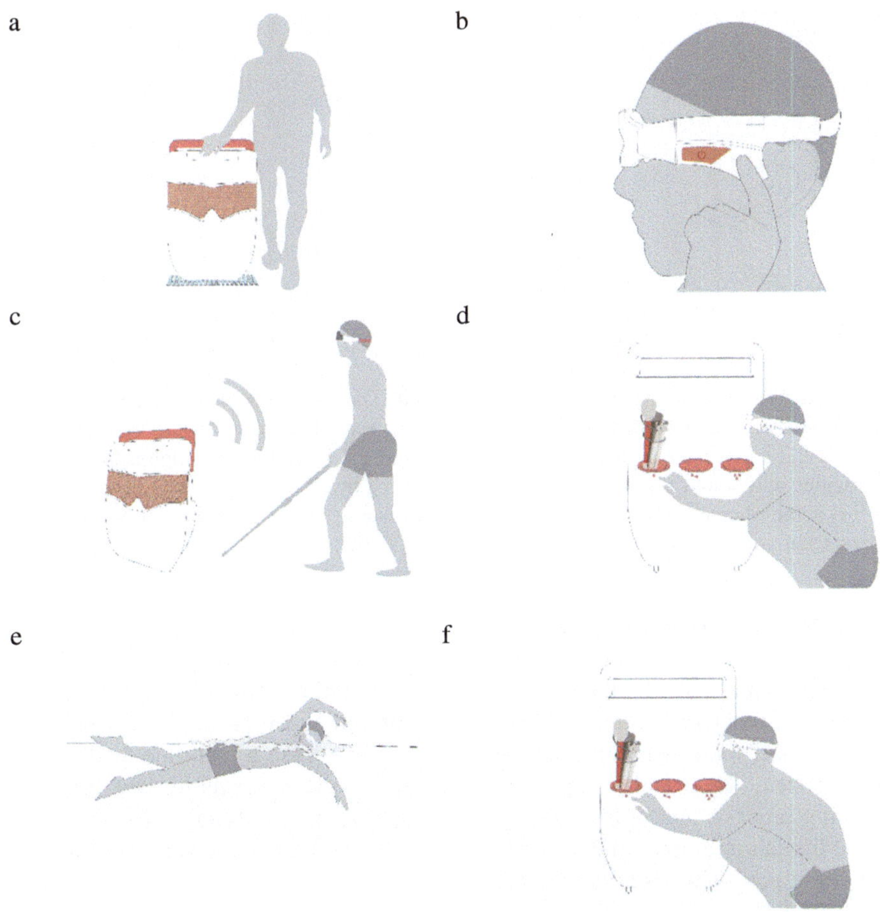

Fig. 11. The procedure of this design: (a) Dealers set up environmental sensor at both ends of the swimming lane. Fix it up and switch on the power. (b) Visually impaired swimmers wear on goggles and turn on the power. (c) With the approach of users, the environmental sensor can further be detected by the headset to send out feedback. (d) Before swim users keep their white cane in the specific slot, touch the Braille to memorize the positioned location. (e) When conditions happen while swimming, the goggles give feedback. (f) After swimming, users take back their white cane according to the Braille.

3.6 Conditions

The condition the user may face are as below (see Fig. 12).

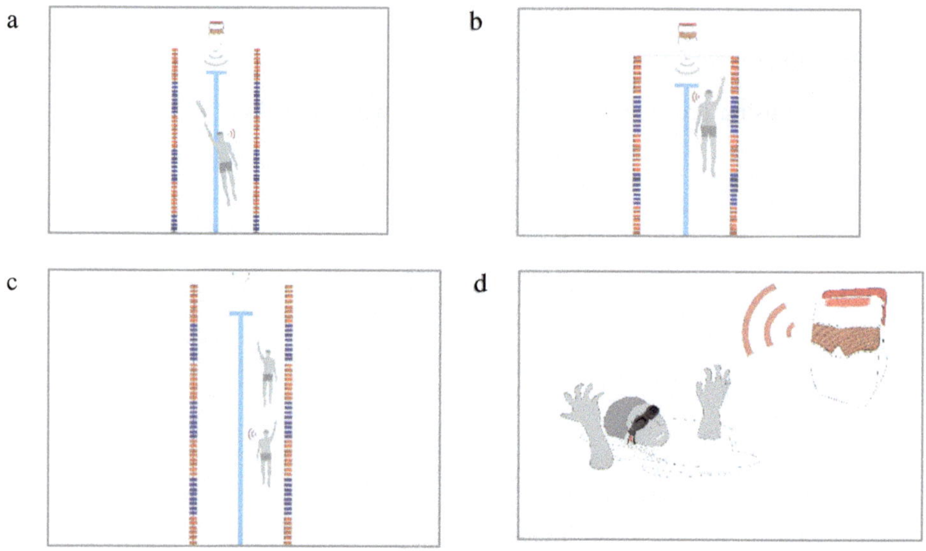

Fig. 12. (a) When the user deviating from right path. The goggles will sound the voice: "Beep! Please turn left/right." to notify the user. (b) Prevent user bumping into wall. The goggles will sound the voice: "Beep! 2 m to wall", to notify the user. (c) When the user approaching someone. The goggles will sound the voice: "Beep! Please slow down", to notify the user. (d) When the user in some danger situation, for example drowning, it will detect abnormal oxygen concentration automatically and send an emergency signal.

4 Discussion and Conclusion

The results of this study let the researcher to understand the difficulties that the visually impaired swimmer had faced and provide the design specification for the researcher. After organize, the researcher uses these results to develop a real-time feedback navigation system for visually impaired swimmers, assist them during the completely swimming process, and provide them a more safety and comfortable situation.

The limitation of this study can be divided into two parts for discussion. Firstly, the observation part. The observation is lack of participant, due to less visually impaired swimmers in Taiwan. From the results of unstandardized interview, we know that a lot of visually impaired in Taiwan nowadays are lack of exercise. There is some reason why they do not exercise. Visually impaired afraid that they will injured during exercise, many exercises need space or site, for example, football need a field, basketball need basketball court. In Taiwan, many of these sites are not friendly for visually impaired, they face obstacle during exercise in those sites and spaces. Another reason is that, some of the sports and activities required special design for visually impaired, guidance or a participant need to assist them during the activities. For example, the visually impaired marathon, another sighted caregiver tied with a visually impaired during the running process.

For the design part, the limitation is lack of test. The design is still a prototype and hard to reach fully waterproof to provide swimmers actual test in a swimming pool. IPX6 [12] are commonly found in other electronics swimming products. IPX6 are highly required in this study to prevent the internal chip from falling when soak into water. To reach IPX6 we need to do a lot of testing and produce generation of fine model for test. Due to this is a one-year project, it is hard to achieved.

Technology limitation. We try to use Wi-Fi to connect between sensor and goggles under the water, but the wave of Wi-Fi interferes by the density of water. Among all the connection, Wi-Fi is the fastest and strongest one, but it spends a lot of battery during the data dissemination process. With this battery lifetime limitation, the goggles may work only for one to two hours.

For further study, the researcher should find more participant. Not only for Taiwan visually impaired swimmers, also the visually impaired swimmers from other country. Increase the visually impaired swimmers as participant may obtain more reliable and precise results.

Due to the unusual or incorrect swimming posture may happen on visually impaired swimmers, a motion capture should be used on the further study. For example, Kinect [13]. Capture the motion of the visually impaired swimmers, compare the posture between visually impaired swimmers and sighted swimmers, maybe that is the good point of outset, to develop a design.

For design, the outlook of the design can be simpler and convenient. The sensor's volume should be reducing and more lightweight.

References

1. WHO: Visual impairment and blindness, 2 February 2016. World Health Organization. http://www.who.int/mediacentre/factsheets/fs282/en/. Accessed 6 Dec 2016
2. Lieberman, L., McHugh, E.: Health-related fitness of children who are visually impaired. J. Vis. Impairment Blind. (JVIB) **95**(5), 272–288 (2001)
3. Houwen, S., Hartman, E., Visscher, C.: Physical activity and motor skills in children with and without visual impairments. Med. Sci. Sports Exerc. **41**(1), 103–109 (2009)
4. Lieberman, L.: Fitness for individuals who are visually impaired or deafblind. RE View **34** (1), 13 (2002)
5. The ASA, the essential element. inclusion of swimmers with a disability. http://www.swimming.org/library/documents/477/download
6. Lieberman, L., McHugh, E.: Health-related fitness of children who are visually impaired. J. Visual Impairment Blind. **95**, 272–287 (2001)
7. Non-participant Observation: A dictionary of sociology. Encyclopedia.com. http://www.encyclopedia.com/social-sciences/dictionaries-thesauruses-pictures-and-press-releases/non-participant-observation. Accessed 8 Jan 2017
8. The Economist: The Hawthorne effect, 3 November 2008. http://www.economist.com/node/12510632. Accessed 9 Jan 2017

9. Fox, N.S., Brennan, J.S., Chasen, S.T.: Clinical estimation of fetal weight and the Hawthorne effect. Eur. J. Obstet. Gynecol. Reprod. Biol. **141**(2), 111–114 (2008). doi:10.1016/j.ejogrb.2008.07.023
10. Holt, R., Thorpe, R. (eds.): The SAGE Dictionary of Qualitative Management Research, p. 117. Sage Publications, London (2007). ISBN 1849203415
11. How bone conduction works. (n.d.). http://www.oticonmedical.com/Medical/YourTreatment/About%20bone%20conduction/How%20does%20it%20work.aspx. Accessed 9 Mar 2017
12. Taylor: Water proof, water resistant, and IPX - what does it all mean? - outdoor tech Blog, 2 July 2013. Science. http://news.outdoortechnology.com/2013/07/02/water-proof-water-resistant-and-ipx-what-does-it-all-mean/Accessed 18 Jan 2017
13. Kinect: Kinect (2009). http://www.xbox.com/zh-TW/xbox-one/accessories/kinect-for-xbox-one

Author Index

T. Ahram (ed.), *Advances in Human Factors in Sports, Injury Prevention and Outdoor Recreation*, Advances in Intelligent Systems and Computing 603,
DOI 10.1007/978-3-319-60822-8

MIX
Papier aus verantwortungsvollen Quellen
Paper from responsible sources
FSC® C105338

If you have any concerns about our products,
you can contact us on
ProductSafety@springernature.com

In case Publisher is established outside the EU,
the EU authorized representative is:
Springer Nature Customer Service Center GmbH
Europaplatz 3, 69115 Heidelberg, Germany

Printed by Libri Plureos GmbH
in Hamburg, Germany